D1800772

Chiral Dynamics

Documents on Modern Physics

Edited by

ELLIOT W. MONTROLL, *University of Rochester*
GEORGE H. VINEYARD, *Brookhaven National Laboratory*
MAURICE LÉVY, *Université de Paris*

A. ABRAGAM L'Effet Mössbauer

S. T. BELYAEV Collective Excitations in Nuclei

P. G. BERGMANN and A. YASPAN Physics of Sound in the Sea: Part I Transmission

T. A. BRODY Symbol-manipulation Techniques for Physics

K. G. BUDDEN Lectures on Magnetoionic Theory

J. W. CHAMBERLAIN Motion of Charged Particles in the Earth's Magnetic Field

S. CHAPMAN Solar Plasma, Geomagnetism, and Aurora

H.-Y. CHIU Neutrino Astrophysics

A. H. COTTRELL Theory of Crystal Dislocations

J. DANON Lectures on the Mössbauer Effect

B. S. DEWITT Dynamical Theory of Groups and Fields

R. H. DICKE The Theoretical Significance of Experimental Relativity

D. FLAMM The Quark Model of Elementary Particles

P. FONG Statistical Theory of Nuclear Fission

E. GERJUOY, B. YASPAN and J. K. MAJOR Physics of Sound in the Sea: Parts II and III Reverberation, and Reflection of Sound from Submarines and Surface Vessels

M. GOURDIN Lagrangian Formalism and Symmetry Laws

D. HESTENES Space–Time Algebra

J. G. KIRKWOOD Dielectrics—Intermolecular Forces—Optical Rotation

J. G. KIRKWOOD Macromolecules

J. G. KIRKWOOD Proteins

J. G. KIRKWOOD Quantum Statistics and Cooperative Phenomena

J. G. KIRKWOOD Selected Topics in Statistical Mechanics

J. G. KIRKWOOD Shock and Detonation Waves

J. G. KIRKWOOD Theory of Liquids

J. G. KIRKWOOD Theory of Solutions

V. KOURGANOFF Introduction to the General Theory of Particle Transfer

R. LATTÈS Methods of Resolution for Selected Boundary Problems in Mathematical Physics

B. W. LEE Chiral Dynamics

J. LEQUEUX Structure and Evolution of Galaxies

J. L. LOPES Lectures on Symmetries

F. E. LOW Symmetries and Elementary Particles

A. MARTIN and F. CHEUNG Analyticity Properties and Bounds of the Scattering Amplitudes

P. H. E. MEIJER Quantum Statistical Mechanics

C. W. MISNER Gravitational Collapse

M. MOSHINSKY Group Theory and the Many-body Problem

M. MOSHINSKY The Harmonic Oscillator in Modern Physics: From Atoms to Quarks

M. NIKOLIĆ Analysis of Scattering and Decay

M. NIKOLIĆ Kinematics and Multiparticle Systems

J. R. OPPENHEIMER Lectures on Electrodynamics

A. B. PIPPARD The Dynamics of Conduction Electrons

H. REEVES Stellar Evolution and Nucleosynthesis

L. SCHWARTZ Application of Distributions to the Theory of Elementary Particles in Quantum Mechanics

J. SCHWINGER Particles and Sources

J. SCHWINGER and D. S. SAXON Discontinuities in Waveguides

M. TINKHAM Superconductivity

J. VANIER Basic Theory of Lasers and Masers

R. WILDT Physics of Sound in the Sea: Part IV Acoustic Properties of Wakes

Chiral Dynamics

BENJAMIN W. LEE

Institute for Theoretical Physics
State University of New York
Stony Brook, New York

QC
174.5
L4

GORDON AND BREACH SCIENCE PUBLISHERS

New York London Paris

WLS 11/20/72

WITHDRAWN
OHIO UNIVERSITY
LIBRARY

Copyright © 1972 by

Gordon and Breach, Science Publishers, Inc.
440 Park Avenue South
New York, N.Y. 10016

Editorial office for the United Kingdom

Gordon and Breach, Science Publishers Ltd.
42 William N St.
London, W.C.2

Editorial office for France

Gordon & Breach
7–9 rue Emile Dubois
Paris 14[e]

This book is the text of a series of lectures given at l'Institut d'Etudes Scientifiques de Cargèse in July, 1970. Supported in part by the AEC Contract AT(30-1) 3668B.

Library of Congress catalog card number 76–146445. ISBN 0 677 01380 9 *(cloth)*; 0 677 01385 X *(paper)*. All rights reserved. No part of this book may be reproduced or utilized in any form or by any means, electronic or mechanical, including photocopying, recording, or by any information storage and retrieval system, without permission in writing from the publishers. Printed in east Germany.

Preface

When Dr. Daniel Bessis asked me to lecture at the Cargèse Summer Institute on chiral symmetry and the σ-model, I was not certain of the level of the audience as they come from all over the world with diverse backgrounds. Since I was planning to give a special topics course on a similar subject during the academic year 1970–71 at Stony Brook to a group of third year graduate students who had taken a two-semester sequence of field theory with me during the preceeding year, I chose to prepare a set of lecture notes aimed at the latter group of students. Thus the present notes presume a knowledge of field theory acquired through a study of, say, the standard textbook on field theory by Bjorken and Drell.

I have made some effort to conform to the conventions of Bjorken and Drell as to the metric, propagators, and normalizations of field operators, etc. A notable exception is in the normalization of single particle states, where I use the convention

$$<p, \alpha|p', \alpha'> = (2\pi)^3 \cdot 2p_0\, \delta^3(p - p')\delta_{\alpha\alpha'}$$

where α and α' stand for quantum numbers such as helicity, isospin, etc.

The present notes are by no means a review of the subject. It is not intended to be. For those desiring a bird's eye view of chiral dynamics, there is already an excellent review by Gasiorowicz and Geffin. Rather, this volume is an intensely personal account of chiral dynamics as I understand it today. In 1967, I have stated that:

> "An optimistic point of view may be that the "success" of the chiral algebra and dynamics is a manifestation of certain characteristics of strong interactions, and that we are beginning to unravel, if not to explain, new principles which govern the subparticle world. On the other hand, it may be argued that we have learned only about the way particles respond to external disturbances (i.e., through electromagnetic and weak interactions), but not much about the

> way particles interact among themselves (through strong interactions), except perhaps the fact that the pion mass is abnormally small in the scale of the subparticle world, for whatever the reason. We do not wish to go farther into this kind of speculations, but must ask you to form your own impression",

with a deliberate ambiguity about the origin of chiral symmetry. Now I feel that I can state with conviction my view on the origin of the chiral SU(2) × SU(2) symmetry. It is that the world is not far away from the Goldstone mode of the chiral SU(2) × SU(2) symmetry as first envisaged by Nambu, and as stressed forcefully by Dashen recently.

I feel less confident about stating my speculation on the way the chiral SU(2) × SU(2) merges with the SU(3) symmetry of hadrons. The pattern of the SU(3) × SU(3) breaking as deduced by Gell-Mann, Oakes, and Renner may have to be further supplemented by a term of non-electromagnetic origin which breaks the isospin conservation to explain the η decays and electromagnetic mass differences. The small parameters appearing in this scheme appear to be of the order of magnitude of the Cabibbo angle. Is there any connection between this pattern and the Cabibbo angle? Despite the valiant efforts by Cabibbo and his collaborators and others, I do not think that the last word on this subject has yet been heard. As for the chiral dynamics applied to K and η mesons, we have not had much success, but have had a few embarrassments. No doubt in time we shall overcome. But to me this is where we stand.

On this occasion I wish to express my gratitude to a number of friends, teachers and colleagues for my own education in this fascinating field, though none of them are responsible for the shortcomings of my thoughts and presentation, of course. C. N. Yang and M. Dresden have provided me, in the last four years, with a shelter where I can do my thing. J. Schwinger's stay in Stony Brook was very helpful in my coming to grips with the phenomenological approach to particles and fields. B. Zumino and K. Symanzik were most kind in sharing with me their unpublished ideas during my sojourn in Europe in the academic year 1968–69. S. Weinberg and R. Dashen, through personal contacts and through their papers, have influenced my thinking profoundly. M. Lévy has been instrumental in arranging the publication of this monograph as a volume in the series, "Documents on Modern Physics". I am grateful to the staff of the Stony Brook Institute, especially to Miss P. Bourie and Mrs. B. Denham for the painstaking typing of the manuscript. I owe much to J. S. Kang for his conscientious assistance in the preparation of the manuscript.

References

J. D. Bjorken and S. Drell, Relativistic Quantum Fields (McGraw-Hill, 1965).

W. A. Bardeen and B. W. Lee, "Nuclear and Particle Physics" edited by B. Margolis and C. S. Lam (Gordon and Breach 1967).

S. Gasiorowicz and D. A. Geffin, *Revs. Modern Phys.* **41** 531 (1969).

Stony Brook, September 1970

Notes added in proof

Looking back at this monograph a year after it was written, I feel that I would have done quite differently on some chapters, were I to write it today. For example, I would have paid more attention to the asymptotic nature of the expansion of the T-matrix in powers of ξ^2 and ε (chapter VI), and would have included a discussion of the so-called σ term in the pion nucleon scattering amplitude (Sec. 10c). There are other shortcomings of the manuscript which became painfully clear since the time of writing. But the present form must remain more or less intact in order to avoid further delay in publication.

I have taken advantage of the informality of these notes to indulge in a certain amount of carelessness. I must emulate George Mackey and tell the reader that if he thinks a sign should be changed, or an i should be inserted, he is probably right. I have made no effort to update the bibliography. Those papers that are cited are those I consulted for preparation of the lectures.

Last, but not least, I must thank M. C. for her constant understanding, and for her considerable help in proofreading.

Stony Brook, November 1971.

Contents

Chiral dynamics

BENJAMIN W. LEE

Institute for Theoretical Physics
State University of New York
Stony Brook, New York

I Hypothesis of partially conserved axial vector currents

Historically, the hypothesis of partially conserved axial vector currents has its essential genesis in the remarkable Goldberger-Treiman relation, which relates the pion decay constant f_π to the nucleon mass and the strong pion-nucleon coupling constant. We need not go into the original derivation of the relation, but shall merely note that the concept of *PCAC* was first introduced to provide a sound foundation for this relation.

Let us first note that the charged pions do decay into the $\mu\nu$ pairs, and according to our understanding of weak interactions, it means that the isospin-carrying axial vector current $\mathbf{A}_\mu(x)$ has a nonvanishing expectation value between the vacuum and the one pion state:

$$\langle 0 | \mathbf{A}_\mu(x) | \pi^k(q) \rangle = i f_\pi q_\mu \boldsymbol{\varepsilon}^k e^{-iq\cdot x}, \tag{1}$$

where k is the isospin index of the pion, and f_π is the pion decay constant, by definition. Experimentally $f_\pi \simeq 95$ Mev as deduced from the known $\pi^\pm \to (\mu^\pm \nu)$ decay rate. Taking the divergence of Eq. (1), we find that

$$\partial^\mu \langle 0 | \mathbf{A}_\mu(x) | \pi^i(q) \rangle = f_\pi m_\pi^2 \boldsymbol{\varepsilon}^i e^{-iq\cdot x}. \tag{2}$$

From Eq. (2), the following conclusions follow immediately:

1) the axial vector currents $\mathbf{A}_\mu(x)$ are not conserved, since neither f_π nor m_π^2 is equal to zero.

2) in the *LSZ* formalism of quantum field theory, the divergence of the axial vector current $\partial^\mu A_\mu^i(x)$ may be used as an interpolating field

for the pion $\pi^i(x)$. The meaning of *PCAC*, as we understand it today, is that the actual world is not far away from the limit in which the axial vector currents are conserved at the expense of having zero mass bosons (i.e. $m_\pi^2 = 0, f_\pi \neq 0$).

Let us make this remark somewhat more precise. Consider the axial charge defined as

$$Q_5^i(t) = \int d^3x\, A_0^i(\mathbf{x}, t),$$

which is in general time dependent. The time rate of change of the axial charge is given by

$$\frac{d}{dt} Q_5^i(t) = \int d^3x\, \partial^\mu A_\mu^i(\mathbf{x}, t)$$

$$= i[H, Q_5^i(t)], \tag{3}$$

where H is the Hamiltonian. We shall write H as

$$H = H_0 + \varepsilon H_1,$$

where H_0 is the part which commutes with $Q_5^i(t)$ and εH_1 the part which does not. The parameter ε is supposed to be a small number. *PCAC* is, then, the hypothesis that, but for the small εH_1, the pion mass would be zero and the local conservation of the axial vector currents would hold, i.e.

$$\partial^\mu A_\mu^i(x) = 0(\varepsilon).$$

We shall now briefly review the derivation of Goldberger-Treiman relation along the line first suggested by Nambu. The matrix element of the axial vector current $A_\mu^i(x)$ between two nucleon states is given by

$$\langle N(p')|\, A_\mu^i(0)\, |N(p)\rangle = \bar{u}(p')\, [\gamma_5\gamma_\mu G_A(t) - \gamma_5 q_\mu H(t)]\, \frac{\tau_i}{2}\, u(p), \tag{4}$$

where $G_A(t)$, $H(t)$ are respectively called the axial vector and induced pseudoscalar form factors, and $q = p - p'$, $t = (p - p')^2$. The value of $G_A(t)$ at zero momentum transfer is G_A, the axial vector coupling constant in the β-decay ($\simeq 1.24$). $H(t)$ may be written as

$$H(t) = \frac{2f_\pi g}{t - m_\pi^2} + \bar{H}(t),$$

where g is the pion-nucleon coupling constant, and $\bar{H}(t)$ is regular at $t = m_\pi^2$. The divergence of Eq. (4) is

$$\langle N(p')|\, \partial^\mu A_\mu^i(0)\, |N(p)\rangle = -i\bar{u}(p')\, D(t)\, \gamma_5 \frac{\tau_i}{2}\, u(p),$$

$$D(t) = 2mG_A(t) - q^2 H(t), \tag{5}$$

where m is the nucleon mass. $D(t)$ has a spectral representation

$$D(t) = -\frac{2f_\pi g m_\pi^2}{t - m_\pi^2} + \bar{D}(t): \quad \bar{D}(t) = \int_{9m_\pi^2}^{\infty} \frac{dt'}{t' - t}\, \Gamma(t'). \tag{6}$$

We equate Eqs. (5) and (6) and then take the limit $t = 0$.

$$2mG_A = 2f_\pi g + \bar{D}(0).$$

If we neglect $\bar{D}(0)$, we obtain the desired relation. The question is how to justify the neglect of $\bar{D}(0)$. To say simply that $D(0)$ is dominated by the nearby pole at $t = m_\pi^2$, because of the small m_π^2, would not do, because the residue at this pole is proportional to m_π^2 [see Eq. (2); the factor m_π^2 on the right hand side is purely of the kinematical origin. Note also that we assume f_π to be of order ε^0. That is to say, even in the zero mass limit, pions are coupled to the axial vector current so that $f_\pi \neq 0$]. We argue instead that both $D(t)$ and its pion pole term vanish in the limit $\varepsilon \to 0$:

$$\lim_{\varepsilon \to 0} \langle N(p')|\, \partial^\mu A_\mu^i(0)\, |N(p)\rangle$$
$$= i\varepsilon\, \langle N(p')|\, [H, A_0^i(0)]\, |N(p)\rangle = 0$$

so $\bar{D}(t)$ must be of order $\varepsilon H_1(\sim m_\pi^2)$, and is negligible compared to the typical strong interaction parameters such as $G_A m \sim m$.

Actually the above demonstration breaks down at $\varepsilon = 0$, since in this limit $D(t)$ vanishes identically. On the other hand $H(t)$ now has a pole at $t = 0$, so $tH(t)$ no longer vanishes at $t = 0$. In fact, as $\varepsilon \to 0$, the Goldberger-Treiman relation becomes exact. To see this clearly, we follow Dashen and Weinstein and write Eq. (5) as

$$2mG_A(t) = tH(t) + D(t). \tag{7}$$

The point of writing in this way is that the pion poles of $H(t)$ and $D(t)$ cancel exactly in the combination $tH(t) + D(t)$ and this combination behaves smoothly in the double limit $\varepsilon \to 0$, and $t \to 0$. We have

$$2mG_A(t) = 2f_\pi g + [t\bar{H}(t) + \bar{D}(t)].$$

Now the limit $t \to 0$ may be taken with impunity inside the square bracket, whether $\varepsilon = 0$ or not. We obtain thereby the desired result:

$$mG_A = f_\pi g + 0(\varepsilon). \tag{8}$$

Bibliography

The notion of *PCAC* was developed in a series of papers by Gell-Mann and collaborators and others:

1. M. Gell-Mann and M. Lévy, *Nuovo Cimento*, **16**, 705 (1960).

2. J. Bernstein, S. Fubini, M. Gell-Mann and W. Thirring, *Nuovo Cimento*, **17**, 757 (1960).

3. Chou Kuang-Chao, *JEPT*, **39**, 703 (1963) [*Soviet Physics*, **12**, 492 (1961)].

The approach we adopted was first expounded by

4. Y. Nambu, *Phys. Rev. Letters*, **4**, 380 (1960)

and recently stressed in

5. R. Dashen and M. Weinstein, *Phys. Rev.*, **183**, 1261 (1969).

II Commutator algebra of currents—chiral $SU(2) \times SU(2)$ as an approximate symmetry

We argued that the reason for the validity of *PCAC* lies in the fact that the axial charges Q_5^i commute approximately with the total Hamiltonian H. We assume that the axial charges Q_5^i, together with the generators Q_i of isospin rotations form the Lie algebra $SU(2) \times SU(2)$: if we form chiral charges $Q_i^\pm = \frac{1}{2}[Q_i \pm Q_i^5]$, they obey the commutation relations

$$[Q_i^\pm, Q_j^\pm] = i\varepsilon_{ijk} Q_k^\pm,$$

$$[Q_i^+, Q_j^-] = 0$$

or

$$[Q_i, Q_j] = i\varepsilon_{ijk} Q_k, \tag{1}$$

$$[Q_i, Q_j^5] = i\varepsilon_{ijk} Q_k^5,$$

$$[Q_i^5, Q_j^5] = i\varepsilon_{ijk} Q_k,$$

where ε_{ijk} is the totally antisymmetric Levi-Civita symbol. The chiral $SU(2) \times SU(2)$ generated by these operators is an approximate symmetry of the Hamiltonian.

In the context of Lagrangian field theory the Lagrangian density $\mathscr{L}$ may be split up into two parts

$$\mathscr{L}[\phi] = \mathscr{L}_0 + \varepsilon\mathscr{L}_1, \tag{2}$$

where $\mathscr{L}_0$ is invariant under a group of continuous transformations:

$$\begin{aligned} \delta\mathscr{L}_0 &= 0, \\ \delta\phi_\alpha &= \theta_i(T^i)_{\alpha\beta}\,\phi_\beta \end{aligned} \tag{3}$$

θ_i being the group parameters and iT_i being an appropriate representation of the i-th generator (suitably ordered) of the group. $\varepsilon\mathscr{L}_1$ is a small noninvariant part of the Lagrangian. Let us further assume that $\mathscr{L}_1$ does not contain any derivatives of fields. Equations of motion satisfied by the ϕ_α (Euler-Lagrange equations) imply that, under local gauge transformations,

$$\partial_\mu \frac{\delta L}{\delta\,\partial_\mu\theta_i(x)} - \frac{\delta L}{\delta\theta_i(x)} = 0; \quad L = \int d^4x\mathscr{L}(x). \tag{4}$$

We define the currents by the expression

$$j^i_\mu(x) = \frac{-\delta L}{\delta\,\partial_\mu\theta_i(x)} = \frac{-\delta L_0}{\delta\,\partial_\mu\theta_i(x)}\,. \tag{5}$$

The corresponding charge Q^i is given by

$$Q^i(t) = \int d^3\mathbf{x}\, j^j_0(\mathbf{x}, t). \tag{6}$$

In the quantized version, Eq. (3) may be written as

$$\delta\phi_\alpha = -i[\theta_i Q^i, \phi_\alpha] = \theta_i(T_i)_{\alpha\beta}\,\phi_\beta. \tag{3'}$$

Equation (4) is equivalent to the statement

$$\partial^\mu j^i_\mu(x) = \frac{-\delta L}{\delta\theta_i(x)} = -\varepsilon\frac{\delta L_1}{\delta\theta_i(x)}$$

or

$$\partial^\mu j^i_\mu = i\varepsilon[Q^i, \mathscr{L}_1] = i\varepsilon[j^i_0(x), L_1]. \tag{7}$$

Now returning to the chiral $SU(2) \times SU(2)$ group, we note that the generators Q^i and Q_5^i are given by

$$Q^i = \int d^3x V_0^i(x), \quad Q_5^i(t) = \int d^3x A_0^i(x), \tag{8}$$

where $V_\mu^i(x)$ and $A_\mu^i(x)$ are respectively the vector and axial vector currents and satisfy the divergence conditions

$$\partial^\mu V_\mu^i(x) = 0, \quad \partial^\mu A_\mu^i(x) = i\varepsilon[Q_5^i, \mathscr{L}_1(x)], \tag{9}$$

since Q^i commute with $\mathscr{L}$, and Q_5^i with $\mathscr{L}_0$. In order for the charges to satisfy the Lie algebra (1) it is sufficient that the currents satisfy the local commutation relations:

$$\begin{aligned} \delta(x_0 - y_0)\,[V_0^i(x), j_0^j(y] &= i\varepsilon^{ijk} j_0^k(x)\,\delta^4(x-y), \\ \delta(x_0 - y_0)\,[V_0^i(x), \mathbf{j}^i(y)] &= i\varepsilon^{ijk}\mathbf{j}^k(x)\,\delta^4(x-y) + \text{s.t.}, \end{aligned} \tag{10}$$

where j_μ^i is either V_μ^i or A_μ^i, and

$$\begin{aligned} \delta(x_0 - y_0)\,[A_0^i(x), A_0^j(y)] &= i\varepsilon^{ijk} V_0^k(x)\,\delta^4(x-y), \\ \delta(x_0 - y_0)\,[A_0^i(x), \mathbf{A}^i(y)] &= i\varepsilon^{ijk}\mathbf{V}^k(x)\,\delta^4(x-y) + \text{s.t.} \end{aligned} \tag{10'}$$

In Eqs. (10) and (10′), "s.t." means the Schwinger term which vanishes upon integration over space, being of the form $\nabla[s(x)\,\delta(x-y)]$, with $s(x)$ in general an operator. Equations (10) and (10′) are Gell-Mann's postulate of the current algebra.

According to Eq. (7), as $\varepsilon \to 0$, i.e. as the Lagrangian becomes invariant under the group, all the currents become divergenceless. There are two ways in which the local conservation of currents manifests itself in nature:

I. the first is that the physical states form a representation basis of the group. In particular the vacuum is invariant under the group.

II. the second possibility is that the physical states form a representation basis of a subgroup of the invariance group of the (classical) Lagrangian. In this case there are zero-mass spin-zero bosons which can be created from the vacuum by the action of currents $\tilde{j}_\mu^i$, the space integrals of whose time components $\tilde{Q}^i$ are not generators of the subgroup. The vacuum states (i.e. the lowest eigenstate of H) is invariant under the subgroup, but not under the full group. (In fact the generators Q^i are not defined in general.)

The second possibility is sometimes referred to as the spontaneous breakdown of the symmetry. The point we wish to stress is that this possibility is a completely legitimate way in which nature reveals the form invariance of the Hamiltonian: the stress should be on "symmetry", rather than on "breakdown". We prefer to call the second possibility the "Goldstone mode" of symmetry manifestation after the person who first discussed this possibility in the context of relativistic quantum theory. The first possibility is the classical one; perhaps we should call it the "Wigner mode".

To illustrate the above remark, we shall consider the matrix element of the axial vector current between nucleon states when the axial vector currents are conserved, i.e. $\partial^\mu A^i_\mu = 0$:

$$\langle N(p')|\, \mathbf{A}_\mu \,|N(p)\rangle = \bar{u}(p')\,\gamma_5 G_A(t)\left[\gamma_\mu - q_\mu \frac{2m}{q^2}\right](\boldsymbol{\tau}/2u)\,(p),$$

where we have eliminated the induced pseudoscalar form factor $H(t)$ using the divergencelessness condition. If uncancelled, the $1/q^2$ dependence of $H(t)$ implies the existence of an isotriplet of pseudoscalar mesons (zero mass pions). There are three cases to consider:

Case 1a $G_A \neq 0,\ m = 0$: In this case there are no Goldstone bosons, the residue of the pole at $q^2 = 0$ being zero. $f_\pi = 0;\ m_\pi^2 \neq 0$.

Case 1b $G_A = 0,\ m \neq 0$: In this case there are no Goldstone bosons, the residue of the pole at $q^2 = 0$ being zero. $f_\pi = 0;\ m_\pi^2 \neq 0$.

Case 2 $G_A \neq 0,\ m \neq 0$: In this case, the Goldberger-Treiman relation is exact and the pions are the Goldstone bosons. $f_\pi \neq 0;\ m_\pi^2 = 0$.

We shall construct models for the three cases cited above, later. For the moment, let it suffice to note that neither case 1a nor 1b comes anywhere close to describing the nature as we observe it.

The hypothesis of *PCAC*, as we interpret it, is the statement about the symmetry of strong interactions. It states that strong interactions are approximately chiral $SU(2) \times SU(2)$ symmetric in the Goldstone mode; the pions would be the Goldstone bosons in the symmetry limit; the departure of the nature from the Goldstone mode of chiral symmetry as measured by the parameter ε is small. This is not the only possible interpretation of *PCAC*. We advocate this above all else, however, because in it we find a natural explanation of the abnormally small pion mass, as well as many successes of the current algebra.

Bibliography

For the current algebra, there is now a handy reference:

1. S. L. Adler and R. F. Dashen, *Current Algebra*, W. A. Benjamin, Inc., New York and Amsterdam (1968).

Of course one should go back to the original source for inspiration:

2. M. Gell-Mann, *Phys. Rev.*, **125**, 1067 (1962).
3. M. Gell-Mann, *Physics*, **1**, 63 (1964).

The Goldstone mode, in the context of relativistic field theory, was first discussed by

4. J. Goldstone, *Nuovo Cimento*, **19**, 155 (1961)

and later elaborated in

5. Y. Nambu and G. Jona-Lasinio, *Phys. Rev.*, **122**, 345 (1961); **124**, 246 (1961).
6. J. Goldstone, A. Salam and S. Weinberg, *Phys. Rev.*, **127**, 965 (1962).
7. S. Bludman and A. Klein, *Phys. Rev.*, **131**, 2363 (1962).

See, for recent discussions,

8. The articles of T. W. Kibble and D. Kastler in *Proceedings of 1967 International Conference on Particles and Fields*, C. R. Hagen *et al*., ed., Interscience, New York (1967)

and

9. G. S. Guralnik, C. R. Hagen, and T. W. Kibble, *Advances in Particle Physics*, Vol. II, R. E. Marshak and R. Cool, ed., *Interscience*, New York (1968).

III The σ-model

3a Tensor analysis of $SU(2) \times SU(2)$

Before discussing models which exhibit the Goldstone mode of the chiral $SU(2) \times SU(2)$ symmetry, a brief review of the $SU(2) \times SU(2)$ group is in order.

We first note that the Lie algebra of $SU(2) \times SU(2)$ is isomorphic to the algebra of the 4-dimensional rotation group $R(4)$. Let us label the generators of $R(4)$ by L_{ij}, $i = 1, 2, 3, 4$, $L_{ij} = -L_{ji}$. Then the correspondence is

$$L_{ij} \sim \varepsilon_{ijk} Q_k, \qquad i, j, k = 1, 2, 3$$

and

$$L_{4k} \sim Q_k^5. \tag{1}$$

All the representations of $SU(2) \times SU(2)$ can be built up from the tensor products of the spinors χ_α and $\chi_{\dot\alpha}$ which transform like the (Pauli) spinors under the $SU(2)$ generated by Q_i^+ and Q_i^- respectively. We shall parametrize an element of $SU(2) \times SU(2)$ by two three-dimensional vectors $\boldsymbol{\alpha}$ and $\boldsymbol{\beta}$ such that

$$\begin{aligned} U(\boldsymbol{\alpha}, \boldsymbol{\beta}) &= \exp\{-i[\boldsymbol{\alpha} \cdot \mathbf{Q} + \boldsymbol{\beta} \cdot \mathbf{Q}_5]\} \\ &= \exp\{-i[(\boldsymbol{\alpha} + \boldsymbol{\beta}) \cdot \mathbf{Q}^+ + (\boldsymbol{\alpha} - \boldsymbol{\beta}) \cdot \mathbf{Q}^-]\}. \end{aligned} \tag{2}$$

Then, under an infinitesimal transformation, we have

$$\begin{aligned} \chi_\alpha &\to \chi_\alpha + i[(\boldsymbol{\alpha} + \boldsymbol{\beta}) \cdot \boldsymbol{\tau}/2]_{\alpha\beta}\, \chi_\beta, \\ \chi_{\dot\alpha} &\to \chi_{\dot\alpha} + i[(\boldsymbol{\alpha} - \boldsymbol{\beta}) \cdot \boldsymbol{\tau}/2]_{\dot\alpha\dot\beta}\, \chi_{\dot\beta}. \end{aligned} \tag{3}$$

A spinor χ_α of $SU(2)$ and its complex conjugate $\chi^\alpha \equiv (\chi_\alpha)^*$ are equivalent (or transform in the same manner), the precise correspondence being

$$\chi_\alpha \sim (\sigma_2)_{\alpha\beta}\, \chi^\beta. \tag{4}$$

We shall label irreducible representations of $SU(2) \times SU(2)$ by $[j, j']$, meaning that the representation in question transforms like the $(2j + 1)$ dimensional representation of $SU(2)$ generated by Q^+ and like the $(2j' + 1)$ of $SU(2)$ generated by Q^-. The dimensionality of the $[j, j']$ representation is $(2j + 1)(2j' + 1)$.

Of particular interest in our applications is the $[\frac{1}{2}, \frac{1}{2}]$ representation, which is the regular representation of $R(4)$. We shall denote it by $M_\alpha^{\dot\beta}$. It transforms as

$$M_\alpha^{\dot\beta} \to M_\alpha^{\dot\beta} + i\boldsymbol{\alpha} \cdot [\boldsymbol{\tau}/2, M]_\alpha^{\dot\beta} + i\boldsymbol{\beta} \cdot \{\boldsymbol{\tau}/2, M\}_\alpha^{\dot\beta}, \tag{5}$$

where $\{A, B\} = AB + BA$, and $(\tau M)_\alpha^{\dot\beta} = (\tau)_\alpha^\gamma M_\gamma^{\dot\beta}$, etc. If we write

$$M_\alpha^{\dot\beta} = [\sigma + i\boldsymbol{\pi} \cdot \boldsymbol{\tau}]_{\alpha\beta} \equiv [M]_{\alpha\beta}, \tag{6}$$

then Eq. (5) implies the transformation law of $(\sigma, \boldsymbol{\pi})$:

$$\begin{aligned} \sigma &\to \sigma - \boldsymbol{\beta} \cdot \boldsymbol{\pi}, \\ \boldsymbol{\pi} &\to \boldsymbol{\pi} - \boldsymbol{\alpha} \times \boldsymbol{\pi} + \boldsymbol{\beta}\sigma, \end{aligned} \tag{7}$$

which shows clearly the 4-vector nature of $(\sigma, \boldsymbol{\pi})$ under $R(4)$. The quantity $\sigma^2 + \boldsymbol{\pi}^2$ is therefore an invariant under $R(4) \sim SU(2) \times SU(2)$. We shall henceforth label the four components of a four vector by the index 0, 1, 2, 3 so that the first, second and third components form a three vector of the given subgroup $R(3)$ (in our case, isospin group). Note further that

$$M_{\dot{\alpha}}^{\beta} = [\sigma - i\boldsymbol{\pi} \cdot \boldsymbol{\tau}]_{\alpha\beta} \equiv [M^{+}]_{\alpha\beta}. \tag{6}$$

There is an automorphism of $SU(2) \times SU(2)$:

$$Q_{+}^{i} \leftrightarrow Q_{-}^{i}, \tag{8}$$

which may be interpreted as the parity operation

$$P: \quad Q^{i} \rightarrow Q^{i}, \qquad Q_{5}^{i} \rightarrow -Q_{5}^{i}. \tag{9}$$

From Eq. (8) and the transformation properties of M_{α}^{β}, it follows that

$$\mathrm{P:} \quad M_{\alpha}^{\beta}(\mathbf{x}, t) = \eta M_{\dot{\alpha}}^{\beta}(-\mathbf{x}, t), \qquad = \eta[M^{+}(-\mathbf{x}, t)]_{\alpha\beta},$$

where η is an arbitrary phase $= \pm 1$. We see therefore that

$$\mathrm{P:} \quad \sigma(\mathbf{x}, t) \rightarrow \eta\sigma(-\mathbf{x}, t), \qquad \boldsymbol{\pi}(\mathbf{x}, t) \rightarrow -\eta\boldsymbol{\pi}(-\mathbf{x}, t).$$

3b Symmetry of the σ-model

The so-called σ-model, which was devised by Schwinger and further elaborated in the context of *PCAC* by Gell-Mann and Lévy, is defined by the Lagrangian

$$\mathscr{L} = \mathscr{L}_0 + \varepsilon\mathscr{L}_1,$$

$$\mathscr{L}_0 = \bar{\psi}[i\gamma_\mu \partial^\mu - g(\sigma + i\boldsymbol{\pi} \cdot \boldsymbol{\tau}\gamma_5)]\psi + \frac{1}{2}[(\partial_\mu\sigma)^2 + (\partial_\mu\boldsymbol{\pi})^2] - \frac{\mu^2}{2}[\sigma^2 + \boldsymbol{\pi}^2] - \frac{\lambda^2}{4}[\sigma^2 + \boldsymbol{\pi}^2]^2, \tag{1}$$

$$\mathscr{L}_1 = c\sigma.$$

Here ψ in an isodoublet fermion field, and $(\sigma, \boldsymbol{\pi})$ are mesons. The parity of σ is plus, and that of the π's minus. The positivity of the Hamiltonian derived from Eq. (1) demands that $\lambda^2 > 0$. We shall choose, by convention, λ to be a positive number $\lambda > 0$.

If we endow ψ with the transformation law:

$$\tfrac{1}{2}(1 - \gamma_5)\,\psi_\alpha \sim \chi_\alpha,$$

$$\tfrac{1}{2}(1 + \gamma_5)\,\psi_\alpha \sim \chi_{\dot{\alpha}}, \tag{2}$$

that is

$$\psi \to \psi + i \cdot \tfrac{1}{2}\boldsymbol{\alpha} \cdot \boldsymbol{\tau}\psi - i \cdot \tfrac{1}{2}\boldsymbol{\beta} \cdot \boldsymbol{\tau}\gamma_5\psi, \tag{2'}$$

and assign the $(\sigma, \boldsymbol{\pi})$ to the $[\frac{1}{2}, \frac{1}{2}]$ representation of $SU(2) \times SU(2)$, then $\mathscr{L}_0$ is invariant under the chiral $SU(2) \times SU(2)$. Before discussing the physical consequences of the model, let us explore the symmetry aspect of the model. The vector currents are given by

$$V_\mu^i(x) = \frac{-\delta L}{\delta\, \partial^\mu \alpha^i(x)}$$

$$= +\bar{\psi}\gamma_\mu \frac{\tau^i}{2}\psi + \varepsilon^{ijk}\pi^j\, \partial_\mu \pi^k \tag{3}$$

and the axial vector currents by

$$A_\mu^i(x) = \frac{-\delta L}{\delta\, \partial^\mu \beta^i(x)} = -\bar{\psi}\gamma_\mu\gamma_5 \frac{\tau_i}{2}\psi + \pi^i\, \partial_\mu\sigma - \sigma\, \partial_\mu \pi^i. \tag{4}$$

The divergence of V_μ^i is zero, and the divergence of A_μ^i is given by

$$\partial^\mu A_\mu^i(x) = \frac{-\delta L}{\delta\beta^i(x)} = -\varepsilon\, \frac{\delta L_1}{\delta\beta^i(x)} = \varepsilon c \pi^i(x). \tag{5}$$

This is a remarkable equation. We have noted in a previous section that the divergence of the axial vector current, in so far as it has a matrix element connecting the vacuum and a one pion state can be used as an interpolating field for the pion. Equation (5) is in accord with this. Furthermore Eq. (5) states that in the σ-model the divergence of the axial vector current is the canonical pion field in terms of which the Lagrangian is written. Comparing Eq. (1.2) with

$$\langle 0|\, \pi^i(0)\, |\pi^j(q)\rangle = Z_\pi^{\frac{1}{2}}\delta_{ij}, \tag{6}$$

where Z_π is the renormalization constant of the field $\pi^i(x)$, we obtain

$$f_\pi m_\pi^2 = \varepsilon c Z_\pi^{\frac{1}{2}}. \tag{7}$$

A salient feature of the Lagrangian is that, because $\mathscr{L}_1$ is linear in the field σ, it allows the nonvanishing vacuum expectation value of the σ-field. Let v be the vacuum expectation value of σ:

$$\langle\sigma\rangle_0 = v. \tag{8}$$

We may define a new field s by the equation

$$\sigma = s + v \tag{9}$$

so that

$$\langle s\rangle_0 = 0. \tag{10}$$

We shall now rewrite the Lagrangian (1) in terms of the new field s.

$$\mathscr{L} = \mathscr{L}_a + \mathscr{L}_b,$$

$$\begin{aligned}\mathscr{L}_a = \bar{\psi}[i\gamma\cdot\partial - m - g(s + i\boldsymbol{\pi}\cdot\boldsymbol{\tau}\gamma_5)]\,\psi \\ + \tfrac{1}{2}[(\partial\boldsymbol{\pi})^2 - \mu_\pi^2\boldsymbol{\pi}^2] + \tfrac{1}{2}[(\partial s)^2 - \mu_\sigma^2 s^2] \\ - \lambda^2 v s(s^2 + \boldsymbol{\pi}^2) - \frac{\lambda^2}{4}(s^2 + \boldsymbol{\pi}^2)^2,\end{aligned} \tag{11}$$

$$\mathscr{L}_b = (\varepsilon c - v\mu_\pi^2)\, s.$$

We have dropped an inessential c-number constant in the expression given above. We have also used the abbreviations:

$$\begin{aligned} m &= gv, \\ \mu_\pi^2 &= \mu^2 + \lambda^2 v^2, \\ \mu_\sigma^2 &= \mu^2 + 3\lambda^2 v^2. \end{aligned} \tag{12}$$

In the form of Eq. (11), the fermions (which we shall call nucleons) have the bare mass m; the pions and the σ are no longer degenerate in mass.

How does one determine the value of v? Well, it must be determined from the condition (10). In perturbation theory there is the lowest order contribution to the vacuum expectation value $\langle s\rangle_0$ proportional to $(\varepsilon c - v\mu_\pi^2)$, and higher order contributions, also, the sum of which we shall denote by $-S(v)$. The condition (10) implies

$$\varepsilon c - v\mu_\pi^2 - S(v) = 0. \tag{13}$$

Feynman diagrams which contribute to $S(v)$ are the so-called tadpole diagrams:

FIGURE 1 Diagrammatic representation of the eigenvalue Eqs. (3b, 13). The cross represents the contribution of $\mathscr{L}_b$. The Feynman rules used are shown in Fig. 2

An example is the diagram in which the σ of zero four momentum dissociates into a virtual pair of a nucleon and an antinucleon and the pair annihilated into the vacuum. $S(v)$ is a complicated nonlinear function of v: in the example cited, the dependence on v creeps in through the nucleon mass m which is gv. Equation (13) may be solved in principle for v, in terms, for example, of μ^2, g and λ^2. We need not worry about this point at this time, however, since we shall learn to write Eq. (13) in a more sensible and compact way.

In generating the perturbation series from the Lagrangian $\mathscr{L}$ of Eq. (10), we may simply take $\mathscr{L}_a$ and forget about $\mathscr{L}_b$, and at the same time disregard any Feynman diagrams in which any s-lines disappear into the vacuum directly through $\mathscr{L}_b$ or indirectly through tadpoles, since according to Eq. (13), the role of $\mathscr{L}_b$ is precisely to cancel the tadpole diagrams. Thus, in computing the Green's functions from $\mathscr{L}_a$, the parameter v may be regarded as free. Later the value of v may be determined in terms of the fundamental parameters of the theory from the eigenvalue equation (13) or its equivalent.

3c Perturbation scheme

It is of course possible to develop a perturbation scheme based on the Lagrangian in the form of Eq. (3b, 1). We shall however describe a scheme based on the form, Eq. (3b, 11). Both methods are of course completely equivalent. Before discussing the details, we must settle on the parameter in which to expand the Green's functions. The most convenient choice that I know is to regard m, μ_π^2, μ_σ^2, and c as fixed parameters and expand the quantities of interest in λ. We see from Eq. (3b, 12) that

$$v = \left[\frac{\mu_\sigma^2 - \mu_\pi^2}{2}\right]^{\frac{1}{2}} \lambda^{-1} \equiv \alpha\lambda^{-1}: \quad \alpha = \left[\frac{\mu_\sigma^2 - \mu_\pi^2}{2}\right]^{\frac{1}{2}} = \lambda v,$$

$$g = m\left[\frac{2}{\mu_\sigma^2 - \mu_\pi^2}\right]^{\frac{1}{2}} \lambda \equiv \beta\lambda: \quad \beta = m\left[\frac{2}{\mu_\sigma^2 - \mu_\pi^2}\right]^{\frac{1}{2}} \tag{1}$$

$$\mu^2 = (3\mu_\pi^2 - \mu_\sigma^2)/2.$$

With this transformation of parameters, $\mathscr{L}_a$ of Eq. (3b, 11) can be written as

$$\mathscr{L}_a = \mathscr{L}_{(2)} + \mathscr{L}_{(3)} + \mathscr{L}_{(4)},$$

$$\mathscr{L}_{(2)} = \tfrac{1}{2}[(\partial\boldsymbol{\pi})^2 + (\partial s)^2 - \mu_\pi^2\boldsymbol{\pi}^2 - \mu_\sigma^2 s^2] + \bar{\psi}(i\gamma\cdot\partial - m)\psi,$$

$$\mathscr{L}_{(3)} = -\lambda\{\alpha s(s^2 + \boldsymbol{\pi}^2) + \beta\bar{\psi}(s + i\boldsymbol{\pi}\cdot\boldsymbol{\tau}\gamma_5)\psi\}, \tag{2}$$

$$\mathscr{L}_{(4)} = -\frac{\lambda^2}{4}(s^2 + \boldsymbol{\pi}^2)^2$$

and

$$\mathscr{L}_b \equiv \mathscr{L}_{(1)} = s\left(\varepsilon c - \frac{\alpha}{\lambda}\mu_\pi^2\right).$$

Notice that in $\mathscr{L}_a$ the terms quadratic in fields $(s, \boldsymbol{\pi}, \psi)$ are multiplied by λ^0, the terms cubic by λ^1, and the terms quatic by λ^2. This has the following important consequence on the expansion of Green's functions in powers of λ. Consider a diagram consisting of E external lines, I internal lines, V_3 three point vertices and V_4 four point vertices. Since an internal line connects two vertices and an external line is attached to a vertex, we have

$$2I + E = 3V_3 + 4V_4.$$

The number of closed loops L in the diagram is given by

$$L = I + 1 - (V_3 + V_4).$$

If we now combine these two equations, eliminating I, we obtain

$$L = \frac{V_3}{2} + V_4 + 1 - \frac{E}{2}$$

or

$$V_3 + 2V_4 = 2(L - 1) + E.$$

Now, the diagram in question is of order $(\lambda)^{V_3+2V_4}$ since each three point vertex gives rise to one power of λ, and each four point vertex brings two powers of λ. Therefore the diagram is of order $\lambda^{2(L-1)+E}$. This has the following consequences: with our way of counting powers

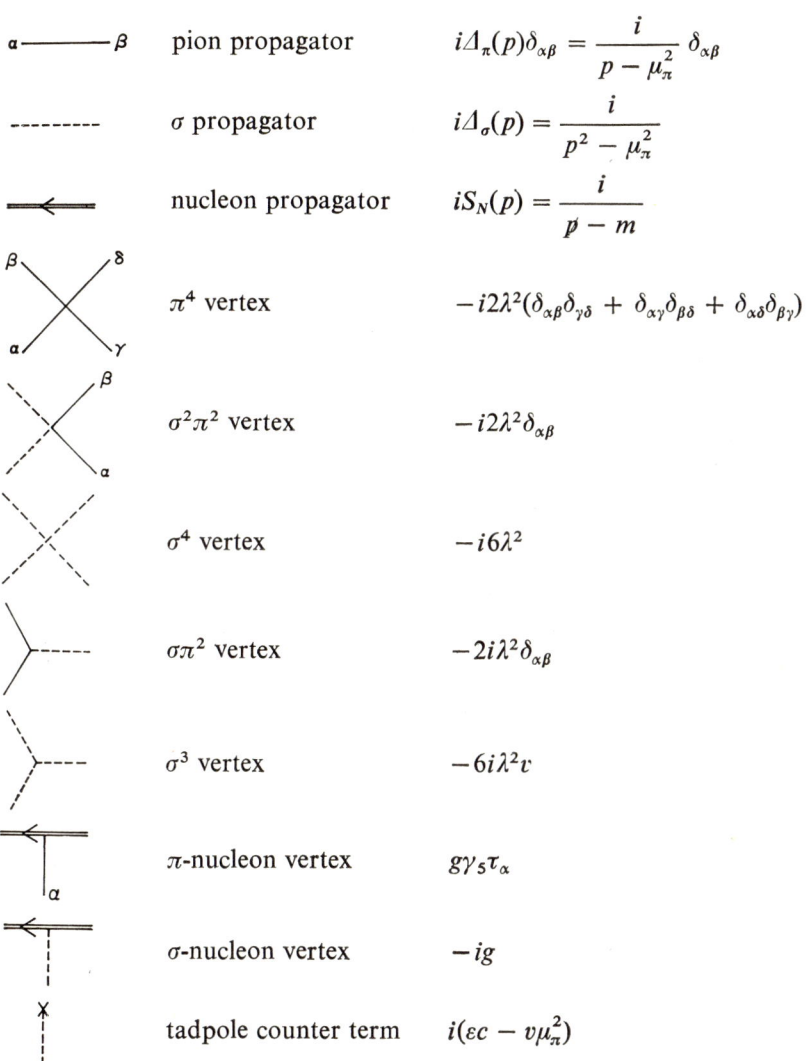

pion propagator	$i\Delta_\pi(p)\delta_{\alpha\beta} = \dfrac{i}{p - \mu_\pi^2}\,\delta_{\alpha\beta}$
σ propagator	$i\Delta_\sigma(p) = \dfrac{i}{p^2 - \mu_\pi^2}$
nucleon propagator	$iS_N(p) = \dfrac{i}{\not{p} - m}$
π^4 vertex	$-i2\lambda^2(\delta_{\alpha\beta}\delta_{\gamma\delta} + \delta_{\alpha\gamma}\delta_{\beta\delta} + \delta_{\alpha\delta}\delta_{\beta\gamma})$
$\sigma^2\pi^2$ vertex	$-i2\lambda^2\delta_{\alpha\beta}$
σ^4 vertex	$-i6\lambda^2$
$\sigma\pi^2$ vertex	$-2i\lambda^2\delta_{\alpha\beta}$
σ^3 vertex	$-6i\lambda^2 v$
π-nucleon vertex	$g\gamma_5\tau_\alpha$
σ-nucleon vertex	$-ig$
tadpole counter term	$i(\varepsilon c - v\mu_\pi^2)$

FIGURE 2 Feynman rules of the σ-model

of λ, the lowest order diagrams with E external lines are of order λ^{E-2} and contain no loops. We call them tree diagrams; the next order diagrams, of order λ^{E} contain one loop each; the diagrams of order λ^{E+2} contain two loops each, etc. In terms of this expansion, Eq. (3b, 13) is of the form

$$\varepsilon c - \frac{\alpha}{\lambda}\mu_{\pi}^{2} - \sum_{n=1}^{\infty} \lambda^{2n-1} f_n(\alpha) = 0,$$

which is a nonlinear eigenvalue equation for $\alpha \equiv \lambda v$. The term $\lambda^{2n-1} f_n(\alpha)$ comes from tadpole diagrams with n loops, since in this case we have $E = 1$.

In a later section (Sec. 5a), we shall show that the consequences of the *PCAC* condition, Eq. (3b, 5), hold in each order of perturbation theory (this is analogous to the situation in quantum electrodynamics, where the consequences of gauge invariance may be verified in each order of perturbation theory).

The Feynman rules are obtained in the usual manner, if we consider $\mathscr{L}_{(2)}$ as the free Lagrangian, and $\mathscr{L}_{(3)} + \mathscr{L}_{(4)}$ as the interaction Lagrangian. The Feynman rules are summarized in Figure 2.

It should also be mentioned that with a diagram with closed loops goes a factor (symmetry number)$^{-1}$. This rule follows directly from the Dyson-Wick expansion of Green's functions. The symmetry number of a diagram is the number of ways in which the same diagram could be drawn if the internal lines of the same kind were regarded as distinct. In enumerating this number, the vertices connected to external lines should be considered as fixed. The vertices which are connected solely to internal lines may be permuted, provided that the same lines meet at a given vertex after the permutation. A few examples of the symmetry number S are shown below.

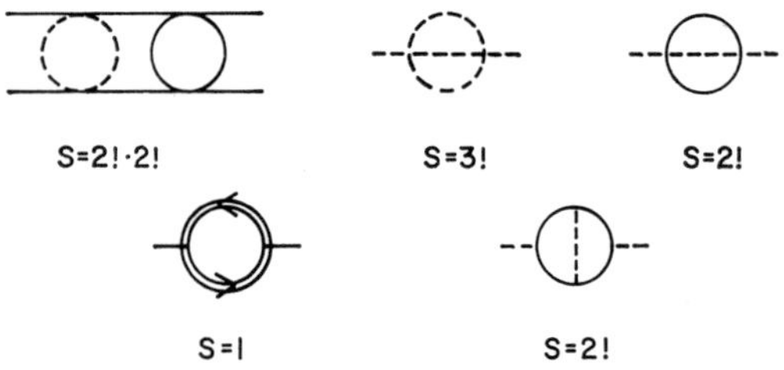

FIGURE 3 Examples of symmetry numbers

The following diagrams are to be multiplied by 1/2:

FIGURE 4 Diagrams which should be multiplied by 1/2

This rule follows, again, from the Dyson-Wick expansion.

3d Tree approximation

We shall examine the physical consequences of the σ-model in the tree approximation. In this approximation the physical masses M, m_π and m_σ of the nucleon, pion and σ are given, respectively, by

$$\begin{aligned} M &= m = gv, \\ m_\pi^2 &= \mu_\pi^2 = \mu^2 + \lambda^2 v^2, \\ m_\sigma^2 &= \mu_\sigma^2 = \mu^2 + 3\lambda^2 v^2. \end{aligned} \tag{1}$$

The value of the vacuum expectation of the σ field is given by the equation

$$\varepsilon c = v m_\pi^2. \tag{2}$$

Recalling that $\varepsilon c = f_\pi m_\pi^2$ [Eq. (3b, 7); to this order $Z_\pi = 1$], we find

$$v = f_\pi, \tag{3}$$

i.e., the pion decay constant is the vacuum expectation value of the σ-field. The axial vector coupling constant G_A is one.

We shall now consider the limit $\varepsilon \to 0$ in which the Lagrangian is exactly chiral symmetric. We see from Eq. (2) that, as $\varepsilon \to 0$, we must have either $v = 0$, or $m_\pi^2 = 0$. Whether one or the other possibility is realized depends on the parameter μ^2. Let us recall that $\lambda^2 > 0$.

1) $\mu^2 > 0$: Since $\lambda^2 v^2 \geqq 0$, $m_\pi^2 \neq 0$. Therefore we must have $v = 0$. In this case, $m_\pi^2 = m_\sigma^2$; $f_\pi = 0$; $m_N = 0$. The symmetry manifests itself in the Wigner mode. This is the case 1a discussed in Sec. 2.

2) $\mu^2 < 0$: In this case $v \neq 0$, since $m_\pi^2 = m_\sigma^2 < 0$ if $v = 0$, and this is physically unacceptable. Therefore we must have $m_\pi^2 = 0$; $f_\pi \neq 0$; $M = gf_\pi$ which is the Goldberger-Treiman relation. The Goldstone mode of symmetry is realized. The symmetry of the vacuum is that of the isospin. This is the case 2 of Sec. 2.

3) $\mu^2 = 0$: We must have $v = 0$; $M = m_\pi = m_\sigma = 0$.

To appreciate better how the different limits are reached as $\varepsilon \to 0$, we shall plot the lines of constant $(\varepsilon c/\lambda^2)$ and constant (μ^2/λ^2) in the two dimensional f_π vs. (m_π^2/λ^2) plane. From Eqs. (1–2), we have

$$f_\pi^2 + \frac{\mu^2}{\lambda^2} = \frac{m_\pi^2}{\lambda^2}$$

$$(\varepsilon c/\lambda^2) = f_\pi(m_\pi^2/\lambda^2).$$

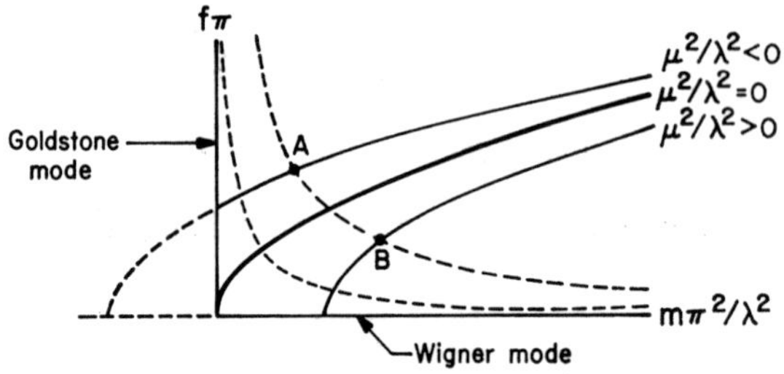

FIGURE 5 The plot of f_π vs. $(m_\pi/\lambda)^2$. The dotted lines correspond to two different values of $(\varepsilon c/\lambda^2)$

The lines of constant μ^2/λ^2 are parabolas and the lines of constant ε hyperbolas. We have plotted the case $f_\pi > 0$. The case $f_\pi < 0$ is obtained by reflection of the plot about the x-axis. If we start from a point on the line $\mu^2/\lambda^2 < 0$ (A in the figure) and let $\varepsilon \to 0$, while keeping μ^2/λ^2 fixed, we will reach a point on the y-axis where $f_\pi \neq 0$, $m_\pi^2 = 0$ (Goldstone mode). On the other hand, if we start from a point on the line $\mu^2/\lambda^2 > 0$, and let $\varepsilon \to 0$, μ^2/λ^2 being kept fixed, we will reach a point on the x-axis on which $f_\pi = 0$, $m_\pi^2 = \mu^2 \neq 0$.

We shall call the region above the line $\mu^2/\lambda^2 = 0$, the Goldstone phase; the region below, the normal phase, of the σ-model. Note however that the line $\mu^2/\lambda^2 = 0$ is *not* a line of singularities of Green's functions, if they are expressed in terms of m, m_π^2, m_σ^2 and λ. As we shall show in a later section [Sec. 4c], each term in the perturbation expansion of the Green's functions is continuous across the line $(\mu/\lambda)^2 = 0$, when the former is expressed in terms of the physical parameters m, m_π^2, m_σ^2 and λ. What is not continuous is the dependence of the physical pion mass m_π^2 on the unphysical parameter μ^2. In the approximation discussed in this section, we have

$$\left(\frac{\varepsilon c}{\lambda^2}\right)^2 = \left(\frac{m_\pi^2}{\lambda^2}\right)^3 - \left(\frac{\mu^2}{\lambda^2}\right)\left(\frac{m_\pi^2}{\lambda^2}\right)^2.$$

The equation is cubic in (m_π^2/λ^2); the physics demands that we choose the branch which gives $m_\pi^2 \geqq 0$. Thus

$$m_\pi^2 > 0, \quad \text{for} \quad \mu^2 > 0, \quad \varepsilon = 0,$$
$$m_\pi^2 = 0, \quad \text{for} \quad \mu^2 < 0, \quad \varepsilon = 0.$$

The discontinuity is shown in the figure below:

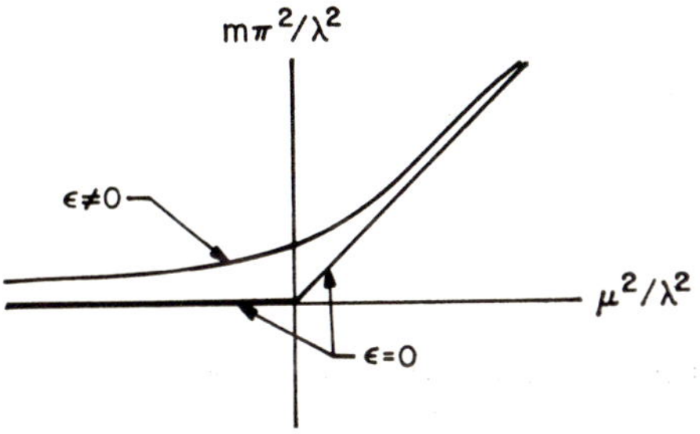

FIGURE 6 The plot of $(m_\pi/\lambda)^2$vs $\cdot$ $(\mu/\lambda)^2$. For $\varepsilon = 0$, the relation is not analytic

The line $\mu^2/\lambda^2 = 0$, as far as I know, has no observable physical significance; I have chosen this line, more or less arbitrarily, as a demarcation line for the modes of symmetry manifestation of the σ-model in the $\varepsilon \to 0$ limit.

The σ-model is found to describe the low energy pion-pion scattering for $\lambda^2 \simeq 5 \sim 6$ [see Sec. 10b, below]. This implies, with

$$f_\pi = 95 \text{ Mev},$$

$$\frac{m_\pi}{\lambda} \cong 56 \text{ Mev},$$

that the nature is in the Goldstone phase.

The foregoing discussion may be understood intuitively if we regard Eq. (3b, 1) as the Lagrangian for classical fields σ and π. The potential energy $V(\sigma, \pi^2)$ is

$$V(\sigma, \pi^2) = \frac{\lambda^2}{4}(\sigma^2 + \pi^2)^2 + \frac{\mu^2}{2}(\sigma^2 + \pi^2) - \varepsilon c\sigma.$$

The vacuum expectation values of the fields are obtained by minimizing the potential energy. We see that the minimum of V is obtained for $\pi = 0$ and $\sigma = v$, where

$$\left.\frac{dV(\sigma, \pi^2)}{d\sigma}\right|_{\sigma=v,\ \pi=0} = \lambda^2 v^3 + \mu^2 v - \varepsilon c = 0.$$

We show the shape of $V(v) = V(v, \mathbf{0})$ for $\mu^2 > 0$ and $\mu^2 < 0$.

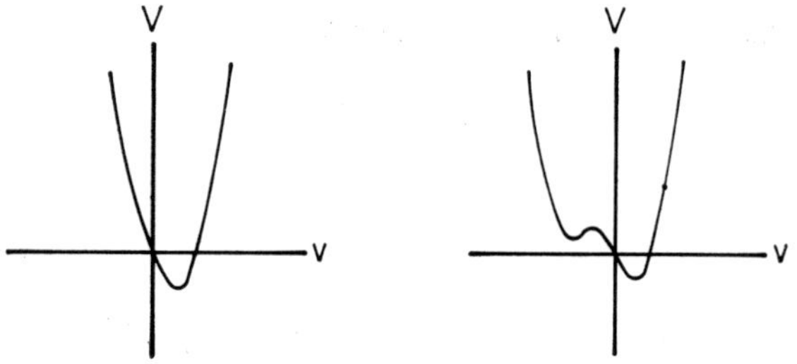

FIGURE 7 The shape of $V(v)$. The right hand side corresponds to $\mu^2 < 0$, the left $\mu^2 > 0$

For $\varepsilon = 0, \mu^2 < 0$, the potential minimum appears on the hypersphere

$$\sigma^2 + \pi^2 = -2\mu^2/\lambda^2$$

and it would appear that, in analogy to the single particle quantum mechanics, the ground state expectation values of the fields should be zero, i.e. $\langle\sigma\rangle_0 = \langle\pi\rangle_0 = 0$. The fact that, in the σ-model, we can localize the "ground state wave function" at the point $\sigma = -2\mu^2/\lambda^2$, $\pi = 0$ on the hypersphere depends in an essential way on the infinite degrees of freedom the system under consideration possesses. For $\varepsilon = 0$, any point on this hypersphere is equivalent, and by an $R(4)$ rotation it may be brought to the standard form $\sigma \neq 0$, $\pi = 0$.

3e Parity doublet model

In Sec. 2, we noted that it is possible to have $G_A = 0$, $m \neq 0$ in the chiral $SU(2) \times SU(2)$ limit. The σ-model of Eq. (3b, 1) does not exhibit this mode of symmetry manifestation. The reason for this is that, in this model, chiral components of the nucleon field transform among themselves [see Eq. (3b, 2)] so that the scalar invariant $\bar{\psi}\psi$ is not a chiral invariant quantity. Theoretically it is possible to assign an alternative transformation law to the isodoublet fermion fields. Assume there exists a pair of isodoublet fermion fields.

$$\Psi = \begin{pmatrix} \psi_1 \\ \psi_2 \end{pmatrix},$$

where ψ_1 and ψ_2 are, each, isospinors. Under the parity operation Ψ transforms as

$$P\colon\ \Psi(\mathbf{x}, t) = \varrho_3 \gamma_0 \Psi(-\mathbf{x}, t),$$

where ϱ_i, $i = 1, 2, 3$ are Pauli matrices acting on the 2×2 parity doublet space. We endow Ψ with the transformation properties

$$\Psi \to \Psi + i \cdot 1(\boldsymbol{\alpha} \cdot \boldsymbol{\tau}/2)\,\Psi + i\varrho_2(\boldsymbol{\beta} \cdot \boldsymbol{\tau}/2)\,\gamma_5 \Psi .$$

There are two invariants under the chiral $SU(2) \times SU(2)$:

$$\bar{\Psi}\Psi,\ \bar{\Psi}\varrho_2\Psi .$$

Noting that $\bar{\Psi}\varrho_1\boldsymbol{\tau}\Psi$ and $\bar{\Psi}\varrho_3\Psi$ transform like $\boldsymbol{\pi}$ and σ, respectively, we can construct a desired Lagrangian:

$$\mathscr{L} = \sum_{i=1}^{2} \bar{\psi}_i(i\gamma \cdot \partial - m)\,\psi_i - g[(\bar{\psi}_1\psi_1 - \bar{\psi}_2\psi_2)\,\sigma + (\bar{\psi}_1\boldsymbol{\tau}\psi_2 + \bar{\psi}_2\boldsymbol{\tau}\psi_1) \cdot \boldsymbol{\pi}]$$
$$+ \frac{1}{2}\,[(\partial\sigma)^2 + (\partial\boldsymbol{\pi})^2] - \frac{\mu^2}{2}\,(\sigma^2 + \boldsymbol{\pi}^2) - \frac{\lambda^2}{4}\,(\sigma^2 + \boldsymbol{\pi}^2)^2 + \varepsilon c\sigma .$$

The axial vector currents are given by

$$\mathbf{A}_\mu(x) = i\bar{\Psi}\varrho_2(\boldsymbol{\tau}/2)\,\gamma_\mu\gamma_5\Psi + \text{meson terms},$$

which satisfies the divergence condition

$$\partial^\mu \mathbf{A}_\mu(x) = \varepsilon c \boldsymbol{\pi}(x).$$

In this model, the symmetry limit is given by $m \neq 0$, $G_A = 0$, $\mu^2 \geqq 0$, which is the case 1b mentioned before. For $\mu^2 < 0$, the $\varepsilon \to 0$ limit gives $G_A = 0$, $m_N \neq 0$, $m_\pi^2 = 0$, $f_\pi \neq 0$, but $g_{\pi NN} = 0$. We dismiss this model as physically uninteresting.

Bibliography

The σ-model was first discussed in

1. J. Schwinger, *Ann. Phys.*, (New York) **2**, 407 (1958).

The definition of the axial vector current in the σ-model is due to

2. J. C. Polkinghorne, *Nuovo Cimento*, **8**, 179 (1958).

The *PCAC* was discussed in great detail in the context of the σ-model in

3. M. Gell-Mann and M. Lévy, *Nuovo Cimento* **16**, 705 (1960)

to which later works on the σ-model, in particular, and the current algebra, in general, owe so much.

IV Renormalization of the σ-model

4a Symmetric renormalization

We have seen in the previous section that the σ-model exhibits the partial chiral symmetry we envisaged in the first two sections. In order to completely explore the implications of the model, we must of course go beyond the tree approximation. The amplitudes obtained in the tree approximation satisfy various constraints imposed on by symmetry considerations, but do not incorporate the full implications of unitarity.

We shall discuss the problem of ultraviolet divergences that arise when we compute higher order terms in the perturbation expansion of Green's functions in a more or less intuitive manner. Since the Lagrangian of the σ-model contains only up to quartic terms in field operators, and the couplings are non-derivative, the ultraviolet divergences in higher order terms can be removed by the addition of a finite number of counter terms and renormalizations of field operators and coupling constants. However, if, in order to cancel divergences in the pion and σ self-masses, we have to add two different counter mass terms, $\delta\mu_\sigma^2 \neq \delta\mu_\pi^2$, or if we end up defining two or more *ad-hoc* parameters to make π^4-, σ^4- and

$\pi^2\sigma^2$-couplings finite, the divergence equation (3b, 5), which was derived from the Lagrangian will not be true in the renormalized version of the theory, since the chiral symmetry will be broken by more than the imput term $\varepsilon c\sigma$. We shall show that this does not happen, and that in fact the divergent parts are strictly chiral symmetric, so that chiral symmetric counter terms and renormalizations eliminate all divergences from the theory. We shall accomplish this by showing the above statement to be true in the normal phase of the σ-model and by arguing that the renormalized Green's functions can be continued without difficulty into the Goldstone phase. A more sophisticated renormalization procedure is due to Symanzik which is based on the general renormalization procedure of Bogoliubov, Parasiuk and Hepp and makes an essential use of the Ward-Takahashi identities to be discussed later. For details, the reader is referred to the lectures of Professor Symanzik at the Cargèse Summer Institute in 1970.

To simplify the matter, let us consider first the Lagrangian which involves only the mesons. We will outline the treatment of the case when the nucleons are included later. We write the Lagrangian in the form,

$$\mathscr{L}_a = \tfrac{1}{2}[(\partial\boldsymbol{\pi})^2 - \mu_\pi^2\boldsymbol{\pi}^2] + \tfrac{1}{2}[(\partial s)^2 - \mu_\sigma^2 s^2]$$

$$- \lambda^2 v s(s^2 + \boldsymbol{\pi}^2) - \frac{\lambda^2}{4}(s^2 + \boldsymbol{\pi}^2)^2, \tag{1}$$

$$\mathscr{L}_b = (\varepsilon c - v\mu_\pi^2)\, s\,.$$

The above Lagrangian should not be regarded as normal ordered. This is so because, had Eq. (1) of Sec. 3b been normal ordered, we would not have been allowed to simply translate the σ field by the c-number constant v, as we did to obtain Eq. (3b, 10). It is therefore best to regard both Eqs. (3b, 1) and (3b, 10) not to have been normal ordered. We shall take it for granted that the symmetric theory in the Wigner mode, i.e., Eq. (1) with $\varepsilon = 0$, $v = 0$ and $\mu^2 > 0$, can be made finite by a redefinition of μ^2, and renormalizations of the one coupling constant λ^2 and the fields $\boldsymbol{\pi}$ and σ.

It is not difficult to see that the superficial degree of divergence of a diagram is at most two in this theory. The superficial degree of divergence D of a diagram with L loops, V_3 three point vertices, V_4 four point vertices and I internal meson lines is given by

$$D = 4L - 2I$$
$$= 4 - E - V_3.$$

The statement is obvious, therefore, for Green's functions for which $E \geqq 2$. For tadpole diagrams, we have $E = 1$. However, it is not difficult to convince oneself that tadpole diagrams with loops contain at least one three point vertex.

Let us consider an arbitrary proper (i.e. single particle irreducible) divergent diagram in the model. It is in general of the form:

$$I = (-i\lambda)^{2V_4}(-i\lambda^2 v)^{V_3}\frac{1}{S}\prod_{a=1}^{L}\int d^4 l_a \prod_i \frac{i}{k_i^2 - \mu_\pi^2}\prod_j \frac{i}{k_j^2 - \mu_\sigma^2} \tag{2}$$

where S is the symmetry number of the diagram; L is the number of loops in the diagram: k_i and k_j are momenta carried by internal π and σ propagators, respectively, and are linear functions of the loop momenta l_a and the external momenta. Let us write the propagators appearing in Eq. (2) as

$$\frac{1}{k^2 - \mu_\pi^2} = \frac{1}{k^2 - \mu^2} + \lambda^2 v^2\left(\frac{1}{k^2 - \mu^2}\right)^2 + (\lambda^2 v^2)^2\left(\frac{1}{k^2 - \mu^2}\right)^2 \frac{1}{k^2 - \mu_\pi^2}$$

and

$$\frac{1}{k^2 - \mu_\sigma^2} = \frac{1}{k^2 - \mu^2} + 3\lambda^2 v^2\left(\frac{1}{k^2 - \mu^2}\right)^2 + (3\lambda^2 v^2)^2\left(\frac{1}{k^2 - \mu^2}\right)^2 \frac{1}{k^2 - \mu_\sigma^2} \tag{3}$$

We substitute expressions (3) into Eq. (2). Then I becomes a sum of terms. In each of these terms we can perform the Wick rotation of the $(l_a)_0$ integration contours since $\mu^2 > 0$ (i.e. since we are dealing with the normal phase).

The terms in which all of $(k_i^2 - \mu_\pi^2)^{-1}$ and $(k_j^2 - \mu_\sigma^2)^{-1}$ are replaced by either the first or second terms of Eq. (3) take the form

$$\lambda^{2V_4}(\lambda^2 v)^{V_3}\frac{1}{S}\prod_{a=1}^{L}\int d^4 l_a \prod_i \frac{(\lambda^2 v^2)^{n_i}}{(k_i^2 - \mu^2)^{1+n_i}}\prod_j \frac{(3\lambda^2 v^2)^{n_j}}{(k_j^2 - \mu^2)^{1+n_j}} \tag{4}$$

where n_i, $n_j = 0$ or 1. In a term in which one or more propagators are replaced by the last terms of Eq. (3), subintegrations routed through these propagators are absolutely convergent, since the integrand goes to zero at least as fast as l_a^{-6} as the integration variable $|l_a|$ (the magnitude of a Euclidean 4 vector) goes to infinity. Since the overall integration was at most quadratically divergent, and, by replacing one or more propagators by the last terms of Eq. (3), the superficial degree of divergences has been lowered at least by four units, the overall integration is superficially convergent in such a term. Other subintegrations may still contain divergences. However, these subintegrations are again of the form of Eq. (4) (but, of course, with different V_3, V_4 and L), and need not be considered separately. In the term in which all propagators are replaced by the last terms of Eq. (3), all subintegrations as well as the overall integration are convergent.

The integral in Eq. (4) is identical to that in the symmetric theory ($\varepsilon \to 0$) in the Wigner mode, in which

$$J = V_3 + 2\Sigma n_i + 2\Sigma n_j \tag{5}$$

number of external σ momenta are put equal to zero. Therefore it is possible to eliminate divergences in Eq. (4) if we choose the counter terms and renormalization constants to be those which render the symmetric theory in the Wigner mode finite. We shall now present a detailed reasoning leading to this conclusion.

Consider I in Eq. (2) as a function of v and expand it in a Taylor series:

$$I(v) = \sum_{J=V_3}^{\infty} \frac{v^J}{J!} I^{(J)} \tag{6}$$

where

$$\begin{aligned} I^{(J)} &= \left[\left(\frac{d}{dv}\right)^J I(v)\right]_{v=0} \\ &= (-i\lambda^2)^{V_3+V_4} \frac{1}{S} \left[\left(\frac{d}{dv}\right)^J v^{V_3} \prod_{a=1}^{L} \int d^4 l_a \right. \\ &\quad \left. \times \prod_i \frac{i}{k_i^2 - \mu^2 - \lambda^2 v^2} \prod_j \frac{i}{k_j^2 - \mu^2 - 3\lambda^2 v^2}\right]_{v=0}. \end{aligned} \tag{7}$$

The differential operator (d/dv) in the square bracket on the right of Eq. (7) can act either on v^{V_3} or on the propagators $(k_i^2 - \mu^2 - \lambda^2 v^2)^{-1}$

or $(k_j^2 - \mu^2 - 3\lambda^2 v^2)^{-1}$. In the latter case, we have

$$\left[\left(\frac{d}{dv}\right)^{2s} \frac{i}{k_i^2 - \mu^2 - \lambda^2 v^2}\right]_{v=0} = \frac{(2s)!}{2^s} \frac{i}{k_i^2 - \mu^2} \left[(-i \cdot 2\lambda^2) \frac{i}{k_i^2 - \mu^2}\right]^s,$$

$$\left[\left(\frac{d}{dv}\right)^{2s} \frac{i}{k_j^2 - \mu^2 - 3\lambda^2 v^2}\right]_{v=0} = \frac{(2s)!}{2^s} \frac{i}{k_j^2 - \mu^2} \left[(-i \cdot 6\lambda^2) \frac{i}{k_j^2 - \mu^2}\right]^s. \tag{8}$$

The right hand side of Eq. (8) is precisely what one gets if one evaluates the amplitude for the emission of s pairs of σ's from a pion (σ) line which carries the initial momentum $k_i(k_j)$ and then let all the momenta of the emitted σ's go to zero.

We can give a diagrammatic interpretation to Eq. (6) in the following way. Let the Feynman diagram corresponding to I be D. We shall attach a fictitious line (call the v-line) at each three point vertex; n_i pairs of v-lines to the i-th pion propagator at n_i distinct points; n_j pairs of v-lines to the j-th σ propagator at n_j distinct points, where V_3, Σn_i and Σn_j must satisfy Eq. (5). There will emerge a number of diagrams with J v-lines. Now replace all propagators in the resulting diagrams by those of the symmetric theory. In each of these diagrams, replace J v-lines by external σ-lines carrying distinct momenta $q_1 \ldots q_J$ in all possible ways that yield distinct Feynman diagrams of the symmetric theory. $I^{(J)}$ is then the limit as $q_1 \ldots q_s$ go to zero of the sum of all such diagrams:

$$I^{(J)} = \lim_{q_1 \cdots q_J \to 0} \sum K^{(J)}(q_1 \ldots q_J) \tag{9}$$

where the summation is over all distinct Feynman amplitudes $K^{(J)}$ of the symmetric theory which is gotten from D by attaching V_3 external σ lines to the three point vertices, and $\frac{1}{2}(J - V_3)$ pairs of σ lines to the internal lines in all possible ways. [The reader should verify that in Eq. (9) the symmetry numbers work out correctly.]

Equation (6) states that if we add counter terms to the Lagrangian (1) and renormalize the coupling constant λ^2 and the wave functions in such a way that all the divergences disappear in the symmetric theory, then the divergences in the σ-model in the normal phase are also eliminated. When the mass counter terms are added to the Lagrangian, Eq. (6) holds true if we interpret $I(v)$ and $I^{(J)}$ to include contributions from the counter terms. We know from the renormalizability of the symmetric model that if we renormalize λ^2 by the substitution

$$\lambda^2 = \lambda_r^2 \frac{Z_\lambda}{Z_M^2} \tag{10}$$

where Z_λ and Z_M are the vertex and wave function renormalization constants of the symmetric theory, then

$$\begin{aligned} &Z_M^{\frac{1}{2}(E+J)} I^{(J)}(\lambda^2), \\ &= Z_M^{\frac{1}{2}(E+J)} I^{(J)}(\lambda_r^2 Z_\lambda Z_M^{-2}) \\ &\equiv I_r^{(J)}(\lambda_r^2) \end{aligned} \tag{11}$$

is a finite function of λ_r^2. Therefore if we renormalize the σ field and its vacuum expectation value v as well as the π field of the σ-model according to

$$\begin{aligned} (\sigma, \pi) &= Z_M^{\frac{1}{2}}(\sigma_r, \pi_r), \\ v = \langle\sigma\rangle_0 &= Z_M^{\frac{1}{2}}\langle\sigma_r\rangle_0 = Z_M^{\frac{1}{2}} v_r \end{aligned} \tag{12}$$

where Z_M is the wave function renormalization constant of the symmetric theory, then the renormalized expression for I:

$$\begin{aligned} I_r(\lambda_r) &\equiv Z_M^{\frac{1}{2}E} I(v, \lambda^2), \\ &= Z_M^{\frac{1}{2}E} I(Z_M^{\frac{1}{2}} v_r, \lambda_r^2 Z_\lambda Z_M^{-2}), \\ &= \sum_{J=V_3}^{\infty} \frac{(v_r)^J}{J!} I_r^{(J)}(\lambda_r^2) \end{aligned} \tag{13}$$

is also finite in terms of λ_r^2 and v_r.

When the nucleon fields are included in the Lagrangian, essentially the same discussion can be given as above. We shall spare the reader a detailed argument which can be found in the literature [Gervais and Lee, *Nuclear Physics* **B 12**, 627 (1969)]. The main point is to note that

$$\left[\left(\frac{d}{dv}\right)^s \frac{i}{\gamma\cdot p - m}\right]_{v=0} = s!\, \frac{i}{\gamma\cdot p}\left[(-igv)\frac{i}{\gamma\cdot p}\right]^s$$

corresponds to the emission of $s\sigma$ particles of zero momentum from a nucleon line. Equation (6) and the interpretation of $I^{(J)}$ in terms of the symmetric theory hold without modification and we can again conclude that if the counter terms that are necessary to make the symmetric theory are added to the Lagrangian of the σ-model and the fields and their vacuum expectation values and the coupling constants are renormalized as in the symmetric theory, the Green's functions of the σ-model are made finite.

In the symmetric theory in the normal mode, the renormalized nucleon mass is zero, as the general argument [see for example, Baker, Johnson and Lee, *Phys. Rev.*, **133**, B 209 (1964)] indicates, and as one can verify in perturbation theory, the infinities in the fermion self mass can be removed by the renormalization of the nucleon field. In the normal phase of the σ-model, the ultraviolet divergences in the nucleon mass are completely eliminated by the renormalization of the coupling constant g, and of v, in the lowest order expression $m = gv$. For details see Mignaco and Remiddi [*Nuclear Phys.*, to be published].

Perhaps a simple illustration is in order at this time. There are three diagrams D_1, D_2 and D_3 for the pion self mass in the one loop approximation (Note that the diagrams D_1 and D_2 are accompanied by a factor 1/2: see the end of Sec. 3c).

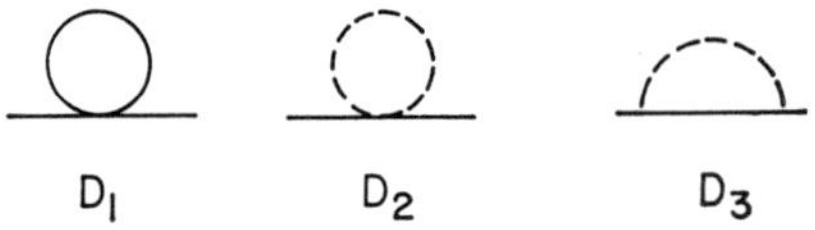

FIGURE 8 Self energy of the pion in the one-loop approximation

When the internal propagators in these diagrams are replaced by the expression (3), there are 5 terms $I^{(1)} \ldots I^{(5)}$ [see the figure below] corresponding to the diagrams $D^{(1)} \ldots D^{(5)}$ which are divergent. In these diagrams, the propagators are those of the symmetric theory.

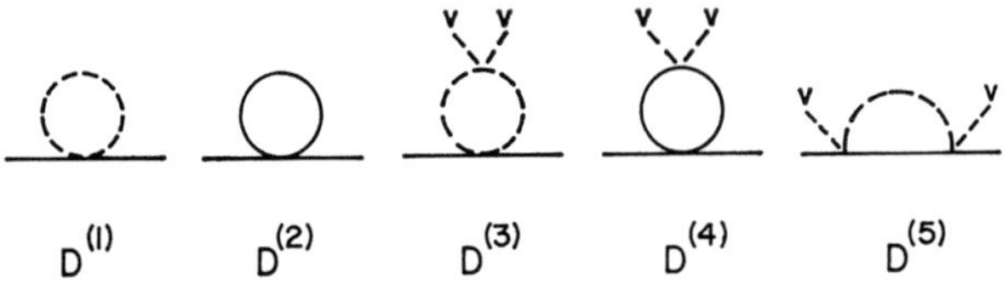

FIGURE 9 Divergent parts of the pion self energy. In these diagrams, the propagators are those of the symmetric theory.The dotted lines with the symbol v are v-lines

The integrals $I^{(1)}$ and $I^{(2)}$ are quadratically divergent and the sum $I^{(1)} + I^{(2)}$ represents the mass-shift of the pion in the symmetric theory. This divergent term, $\delta\mu^2$, should be amalgamated into μ^2 in the expression $\mu_\pi^2 = \mu^2 + \lambda^2 v^2$. The sum $I^{(3)} + I^{(4)} + I^{(5)}$ is equal to $v^2/2!$ times the one-loop correction to the $\pi^2\sigma^2$ vertex in the symmetric theory,

$-2i(\delta\lambda^2)\,\delta_{\alpha\beta}$. Thus the entire divergent contributions from the one-loop diagrams plus the lowest order expression for μ_π^2 can be written as

$$\mu^2 + \delta\mu^2 + (\lambda^2 + \delta\lambda^2)\,v^2. \tag{14}$$

The reader should have no difficulty in verifying that the corresponding expression for σ is

$$\mu^2 + \delta\mu^2 + 3(\lambda^2 + \delta\lambda^2)\,v^2, \tag{15}$$

since the one-loop correction to the σ^4 vertex is of the form $-6i(\delta\lambda^2)$. The parameters $\delta\mu^2$ and $\delta\lambda^2$ appearing in both Eqs. (14) and (15) are the same, and are the mass shift and change in the coupling constant λ^2 in the one-loop approximation in the symmetric theory.

The reader should verify, by the direct evaluation in the one-loop approximation, that the divergences in the $\pi^2\sigma$-, σ^3-, σ^4-, $\sigma^2\pi^2$- and π^4-vertices can be cancelled simultaneously by the redefinition of the coupling constant λ^2 : $\lambda^2 \to \lambda^2 + \delta\lambda^2$ where $\delta\lambda^2$ is the same quantity appearing in Eqs. (14) and (15).

4b Finiteness of the eigenvalue equation and the divergence equation

The preceeding arguments imply that if we write the Lagrangian as

$$\begin{aligned}
\mathscr{L}_a = {} & \bar{\psi}_0[i\gamma\cdot\partial - m_0 - g_0(s_0 + i\boldsymbol{\pi}_0\cdot\boldsymbol{\tau}\gamma_5)]\,\psi_0 \\
& + \tfrac{1}{2}[(\partial s_0)^2 - \mu_{\sigma 0}^2 s_0^2] + \tfrac{1}{2}[(\partial\boldsymbol{\pi}_0)^2 - \mu_{\pi 0}^2\boldsymbol{\pi}_0^2] \\
& - \lambda_0^2 v_0 s_0(s_0^2 + \boldsymbol{\pi}_0^2) \\
& - \frac{\lambda_0^2}{4}(s_0^2 + \boldsymbol{\pi}_0^2)^2,
\end{aligned} \tag{1}$$

$$\begin{aligned}
\mathscr{L}_b &= s_0[\varepsilon c - v_0\mu_{\pi 0}^2], \\
m_0 &= g_0 v_0, \\
\mu_{\pi 0}^2 &= \mu^2 + \delta\mu^2 + \lambda_0^2 v_0^2, \\
\mu_{\sigma 0}^2 &= \mu^2 + \delta\mu^2 + 3\lambda_0^2 v_0^2,
\end{aligned}$$

and renormalized fields and parameters according to the scheme:

$$\begin{aligned}
\psi_0 &= Z_F^{\frac{1}{2}}\psi, \\
(\boldsymbol{\pi}_0, s_0, v_0) &= Z_M^{\frac{1}{2}}(\boldsymbol{\pi}, s, v), \\
g_0 &= gZ_g/Z_F Z_M^{\frac{1}{2}}, \\
\lambda_0^2 &= \lambda^2 Z_\lambda/Z_M^2
\end{aligned} \tag{2}$$

and choose the divergent parts of $\delta\mu^2$, Z_F, Z_M, Z_g and Z_λ to be those that make the symmetric theory in the Wigner mode finite, then the Green's functions of the σ-model become finite in terms of finite parameters μ^2, g, λ^2 and v. [λ_r^2, g_r and v_r of the preceeding subsection are denoted by λ^2, g and v from now on.]

It now remains to show that the eigenvalue equation for v, Eq. (3b, 12), which we shall write as

$$\varepsilon c - v_0(\mu^2 + \delta\mu^2 + \lambda_0^2 v_0^2) - S(v_0) = 0 \tag{3}$$

becomes finite after the renormalization. The precise form of the above equation is obtained from the identity

$$\partial^\mu \langle T(A_\mu^i(x)\, \pi_0^j(0))\rangle_0 = i\delta^{ij}\langle\sigma_0\rangle_0\, \delta^4(x) + \varepsilon c\langle T(\pi_0^i(x)\, \pi_0^j(0))\rangle_0, \tag{4}$$

which gives, upon integrating over all space-time,

$$\varepsilon c = v_0[-\Delta_F^{(\pi)}(0)]^{-1}, \tag{5}$$

where $\Delta_F^{(\pi)}(k^2)$ is the unrenormalized full pion propagator:

$$i\delta^{ij}\, \Delta_F^{(\pi)}(k^2) = \int d^4x\, e^{ik\cdot x}\langle T(\pi_0^i(x)\, \pi_0^j(0))\rangle_0 .$$

Equation (4) is the first of the hierachy of the Ward-Takahashi identities that follow from the divergence equation $\partial^\mu \mathbf{A}_\mu(x) = \varepsilon c \boldsymbol{\pi}_0(x)$. We shall discuss these in the next section.

Comparing Eqs. (3) and (5) we learn that

$$\mu^2 + \delta\mu^2 + \lambda_0^2 v_0^2 + v_0^{-1} S(v_0) = [-\Delta_F^{(\pi)}(0)]^{-1}. \tag{6}$$

If we carry out the renormalization on Eq. (5):

$$v_0 = Z_M^{\frac{1}{2}} v,$$

$$\Delta_F^{(\pi)} = Z_M \Delta_{\text{ren}}^{(\pi)}$$

then it can be written as

$$v[-\Delta_{\text{ren}}^{(\pi)}(0)]^{-1} = \gamma, \tag{7}$$

where

$$\gamma = \varepsilon c Z_M^{\frac{1}{2}}. \tag{8}$$

Note also that with the definition (7), the divergence equation becomes

$$\partial^\mu \mathbf{A}_\mu(x) = \gamma \boldsymbol{\pi}(x). \tag{9}$$

Equations (7) and (9) mean that if the coefficient γ of the renormalized pion field in the divergence equation is finite, the eigenvalue equation for v, Eq. (7), is also finite. Note that $\Delta^{(\pi)}_{\text{ren}}(0)$ is a function of v, μ^2, g and λ^2, for example.

As we have stated repeatedly, the renormalization prescription we gave is meaningful for the normal phase of the σ-model. However, once the Green's functions have been renormalized, they may be expressed in terms of the finite parameters m_π^2, the physical pion mass, g, λ^2 and v, for example, eleminating the unphysical parameter μ^2. Then each term of the perturbation series is well-defined as $m_\pi^2 \to 0$. In particular the Green's functions as functions of m_π^2, g, λ^2 and v are well-behaved as we go from the normal phase to the Goldstone phase along the path $v =$ const., $m_\pi^2 \to 0$. When extended to the Goldstone phase in this manner, Eq. (7) is the statement of the Goldstone theorem: if $\gamma = 0$, then either $v = 0$, or $[\Delta^{(\pi)}_{\text{ren}}(0)]^{-1} = 0$, i.e. $m_\pi^2 = 0$.

4c Renormalization constants

We have stated that the divergent parts of $\delta\mu^2$, Z_M, Z_F, Z_g and Z_λ should be so chosen that they remove the divergences in the symmetric theory in the Wigner mode. How to choose the finite parts of these constants is left largely to one's discretion. The most convenient choice, in terms of constructing the physical T-matrix, is to write the Lagrangian as

$$\begin{aligned}\mathscr{L}_a &= \bar{\psi}_0[i\gamma\cdot\partial - g_0 v_0 - g_0(s_0 + i\gamma_5 \boldsymbol{\pi}_0\cdot\boldsymbol{\tau})]\,\psi_0 \\ &\quad + \tfrac{1}{2}[(\partial\boldsymbol{\pi}_0)^2 - (m_\pi^2+\delta m^2)\boldsymbol{\pi}_0^2 \\ &\quad + \tfrac{1}{2}[(\partial s_0)^2 - (m_\pi^2 + \delta m^2 + 2\lambda_0^2 v_0^2)\, s_0^2] - \lambda_0^2 v_0 s_0 (s_0^2 + \boldsymbol{\pi}_0^2) \\ &\quad - \frac{\lambda_0^2}{4}(s_0^2 + \boldsymbol{\pi}_0^2)^2, \qquad (1)\\ \mathscr{L}_b &= [\varepsilon c - v_0(m_\pi^2 + \delta m^2)]\, s_0\end{aligned}$$

and to renormalize field operators and the constants v, g and λ^2 as in Eq. (4b, 2), and to choose δm^2 so that the physical pion mass is m_π^2, and choose Z_F and Z_M so that

$$\begin{aligned}\langle 0|\, \psi(x)\, |N(p)\rangle &= e^{-ip\cdot x} u(p),\\ \langle 0|\, \pi^i(x)\, |\pi^j(q)\rangle &= e^{-iq\cdot x}\delta^{ji}.\end{aligned}$$

If the σ-meson is stable, the renormalized σ field $\sigma = Z_M^{-\frac{1}{2}}\sigma_0$ is not normalized asymptotically to the unit amplitude:

$$\langle 0|\,\sigma(x)\,|\sigma(q)\rangle \neq e^{-iq\cdot x}.$$

However, for physically interesting applications, the σ meson is always unstable, and does not appear as an asymptotic state. The most convenient renormalization for the σ-field is by Eq. (4b, 2), since in this way the symmetry between the renormalized pion and σ fields are maintained. With this choice of renormalization parameters [see Eq. (4b, 9)], we have

$$\gamma = f_\pi m_\pi^2, \tag{2}$$

so that Eq. (4b, 7) implies

$$f_\pi = v[-\Delta_{\text{ren}}^{(\pi)}(0)]^{-1}/m_\pi^2. \tag{3}$$

Expressed in the form of Eq. (1), the Lagrangian makes no reference to the unphysical mass μ^2. The Feynman integrals contain only the parameters m_π^2, g, λ^2 and v after the renormalization and they do not exhibit any singular behavior at $m_\pi^2 = \lambda^2 v^2$, which corresponds to $\mu^2 = 0$.

The vertex renormalization constants Z_λ and Z_g are to be chosen in accordance with the convention one adopts for the coupling constants λ^2 and g. For example $-2i\lambda^2(\delta_{\alpha\beta}\delta_{\gamma\delta} + \delta_{\alpha\gamma}\delta_{\beta\delta} + \delta_{\alpha\delta}\delta_{\beta\gamma})$ may be defined as the value of the π^4-vertex when all four pion momenta vanish.

Bibliography

The renormalization problem in the σ-model was discussed in

1. B. W. Lee, *Nuclear Physics*, **B 9**, 649 (1969).
2. J.-L. Gervais and B. W. Lee, *Nuclear Physics*, **B 12**, 627 (1969).
3. K. Symanzik, Lettere al *Nuovo Cimento* **1**, 10 (1969).
4. K. Symanzik, *Comm. Math. Phys.*, **16**, 48 (1970).

For a detailed discussion of the renormalization procedure of Symanzik, the reader is referred to the lectures of K. Symanzik at the Cargèse Summer Institute, 1970. A discussion parallel to ours, but making use of the functional integration method popular among Russian authors, was given by A. Vassiliev at the same summer school.

V Ward-Takahashi Identities

5a Connected Green's functions

In this section we shall explore the consequences of the divergence equation of the σ-model,

$$\partial^\mu \mathbf{A}_\mu(x) = \gamma \boldsymbol{\pi}(x) \tag{1}$$

by examining the Green's functions of the form

$$\langle T(A^i_\mu(z)\, s(x_1)\, s(x_2) \dots s(x_n)\, \pi^{i_1}(y_1) \dots \pi^{i_m}(y_m))\rangle_0,$$

where s and π are renormalized fields, $s_0 = Z_M^{\frac{1}{2}} s$, $\boldsymbol{\pi}_0 = Z_M^{\frac{1}{2}} \boldsymbol{\pi}$ so that $\langle 0 \,|\, \pi^i(0) \,|\, \pi^j \rangle = \delta_{ij}$. Taking the divergence of this Green's function and making use of Eq. (1) and

$$\delta(z_0 - x_0)\,[A^i_0(z), s(x)] = -i\pi^i(x)\,\delta^4(x - z), \tag{2}$$

$$\delta(z_0 - y_0)\,[A^i_0(z), \pi^j(y)] = i\delta^{ij}\sigma(y)\,\delta^4(y - z)$$

as follow from the definition of the axial vector currents, Eq. (3b, 4) and canonical commutation rules, we obtain

$$\begin{aligned}
&\partial^\mu \langle T(A^i_\mu(z)\, s(x_1) \dots s(x_n)\, \pi^{i_1}(y_1) \dots \pi^{i_m}(y_m))\rangle_0 \\
&= \gamma \langle T(s(x_1) \dots s(x_n)\, \pi^i(z)\, \pi^{i_1}(y_1) \dots \pi^{i_m}(y_m))\rangle_0 \\
&\quad + i \sum_j \langle T(s(x_1) \dots s(x_n)\, s(z)\, \pi^{i_1}(y_1) \dots \pi^{i_{j-1}}(y_{j-1})\, \pi^{i_{j+1}}(y_{j+1}) \\
&\qquad \dots \pi^{i_m}(y_m))\rangle_0 \times \delta^4(y_j - z)\, \delta_{i,i_j} \\
&\quad + iv \sum_j \langle T(s(x_1) \dots s(x_n)\, \pi^{i_1}(y_1) \dots \pi^{i_{j-1}}(y_{j-1})\, \pi^{i_{j+1}}(y_{j+1}) \\
&\qquad \dots \pi^{i_m}(y_m))\rangle_0 \times \delta^4(y_j - z)\, \delta_{i,i_j} \\
&\quad - i \sum_t \langle T(s(x_1) \dots s(x_{t-1})\, s(x_{t+1}) \dots s(x_n)\, \pi^i(z)\, \pi^{i_1}(y_1) \dots \pi^{i_m}(y_m))\rangle_0 \\
&\quad \times \delta^4(x_t - z).
\end{aligned} \tag{3}$$

We define the Fourier transforms of the Green's functions:

$$\int d^4z\, e^{ik\cdot z} \prod_{i=1}^{n} \int d^4x_i\, e^{ip_i\cdot x_i} \prod_{j=1}^{m} \int d^4y_j\, e^{iq_j\cdot y_j}$$

$$\times \langle T(A^i_\mu(z)\, s(x_1) \ldots s(x_n)\, \pi^{i_1}(y_1) \ldots \pi^{i_m}(y_m))\rangle_0$$

$$\equiv iG^{i;i_1\ldots i_m}_\mu(k; p_1 \ldots p_n; q_1 \ldots q_m)\,(2\pi)^4\, \delta^4(k + \Sigma p_i + \Sigma q_j);$$

$$\prod_{i=1}^{n} \int d^4x_i\, e^{ip_i\cdot x_i} \prod_{j=1}^{m} \int d^4y_j\, e^{iq_j\cdot y_j} \tag{4}$$

$$\times \langle T(s(x_1) \ldots s(x_n)\, \pi^{i_1}(y_1) \ldots \pi^{i_m}(y_m))\rangle_0$$

$$\equiv iG^{i_1\ldots i_m}(p_1 \ldots p_n; q_1 \ldots q_m)\,(2\pi)^4\, \delta^4(\Sigma p_i + \Sigma q_j).$$

[Wow! We must economize in our notation. We will omit all isospin indices and let k, $q_1 \ldots q_m$ stand for isospin indices as well as momenta]. Let k be a space-like vector. Then

$$\int d^4z\, e^{ik\cdot z} \prod_{i} \int d^4x_i\, e^{ip_i\cdot x_i} \prod_{j} \int d^4y_j\, e^{iq_j\cdot y_j}$$

$$\times\ \partial^\mu_z \langle T(A_\mu(z)\, s(x_1) \ldots s(x_n)\, \pi(y_1) \ldots \pi(y_m))\rangle_0 \tag{5}$$

$$= k^\mu G_\mu(k; p_1 \ldots p_n; q_1 \ldots q_m)\,(2\pi)^4\, \delta(k + \Sigma p_i + \Sigma q_j),$$

where we have used Stoke's theorem; the so-called surface term vanishes in as much as k is a space-like vector. Equation (3) can be written in momentum space as

$$k_\mu G^\mu(k; p_1 \ldots p_n; q_1 \ldots q_m) - i\gamma G(p_1 \cdots p_n; kq_1 \ldots q_m)$$

$$= \sum_t G(p_1 \ldots \hat{p}_t \ldots p_n; k + p_t, q_1 \ldots q_m)$$

$$- \sum_s G(k + q_s, p_1 \ldots p_n; q_1 \ldots \hat{q}_s \ldots q_m)\, \delta_{i,i_s}$$

$$- v \sum_s G(p_1 \ldots p_n; q_1 \ldots \hat{q}_s \ldots q_m)\, \delta_{i,i_s} (2\pi)^4\, \delta^4(k + q_s), \tag{6}$$

where the hatted quantities are to be omitted. Equation (6) holds for all n, m except $n = 0$, $m = 1$. For $n = 0$, $m = 1$, we have instead

$$k^\mu G_\mu(k; ;q) - i\gamma G(; kq) = +iv. \tag{7}$$

We may now remove from both sides of Eqs. (6) and (7) the contributions from disconnected graphs. The last term on the right hand side of Eq. (6) corresponds entirely to disconnected graphs. Let us denote by H_μ and H the connected parts of the amplitudes G_μ and G defined in Eq. (3). Then we have

$$\begin{aligned} k^\mu H_\mu(k; p_1 \dots p_n; q_1 \dots q_m) &- i\gamma H(p_1 \dots p_n; kq_1 \dots q_m) \\ = \sum_t H(p_1 &\dots \hat{p}_t \dots p_n; k + p_t, q_1 \dots q_m) \\ - \sum_s H(p_1 &\dots p_n, k + q_s; q_1 \dots \hat{q}_s \dots q_m) \delta_{i, i_s} \end{aligned} \tag{8}$$

for all $n \geqq 0$, $m \geqq 1$ except $n = 0$, $m = 1$ and

$$k^\mu H_\mu(k; ;q) - i\gamma H(; kq) = +iv \tag{9}$$

for $\quad n = 0, \quad m = 1.$

Let us now consider the limit $k \to 0$ in Eqs. (8) and (9). If $\gamma \neq 0$, then the pion mass is finite and as $k \to 0$ the term $k^\mu H_\mu$ goes to zero. On the other hand, if $\gamma = 0$, the second term on the left is zero but $k^\mu H_\mu$, because of the pion pole in H_μ at $k^2 = 0$, goes to a constant. In fact, the limit $k \to 0$ of the left hand side of Eq. (8) is smooth in γ (we have demonstrated the same phenomenon in deriving the Goldberger-Treiman relation in Sec. 1). Let us isolate the pion pole term in H_μ by writing

$$H_\mu(k; p_1 \dots p_n; q_1 \dots q_m) = if_\pi k_\mu H(p_1 \dots p_n; kq_1 \dots q_m) + \bar{H}_\mu,$$

where the pion pole contribution to H_μ is contained in the first term on the right and $\bar{H}$ is nonsingular at $k^2 = m_\pi^2$. The left hand side of Eq. (8) is

$$if_\pi(k^2 - m_\pi^2)\, H(p_1 \dots p_n; kq_1 \dots q_m) + k^\mu \bar{H}_\mu,$$

where we have used the previous result $\gamma = f_\pi m_\pi^2$. As $k \to 0$, the second term vanishes, and the first term goes to a limit whether or not $m_\pi^2 = 0$, since H has a pole at $k^2 = m_\pi^2$. We therefore write

$$\begin{aligned} -i\gamma H(p_1 &\dots p_n; 0q_1 \dots q_m) \\ = \sum_t H(p_1 &\dots \hat{p}_t \dots p_n; p_t q_1 \dots q_m) \\ - \sum_s H(p_1 &\dots p_n q_s, q_1 \dots \hat{q}_s \dots q_m) \end{aligned} \tag{10}$$

for all $n \geqq 0$, $m \geqq 1$ except $n = 0$, $m = 1$

and

$$-\gamma\,\Delta_\pi(0) = v \quad \text{for} \quad n = 0, \qquad m = 1. \tag{11}$$

In Eq. (11) we have used the fact that

$$\begin{aligned} iH(;k,q) &= \int d^4x\, e^{ik\cdot x}\langle T(\pi(x)\,\pi(0))\rangle_0 \\ &\equiv i\,\Delta_\pi(k^2). \end{aligned} \tag{12}$$

Equations (10) and (11) are the fundamental identities upon which our discussion will be based; these are the Ward-Takahashi identities that follow from the divergence equation (1).

Combining Eqs. (10) and (11) we write

$$\begin{aligned} &-v[i\,\Delta_\pi(0)]^{-1}\, H(p_1 \dots p_n; 0q_1 \dots q_m) \\ &= \sum_t H(p_1 \dots \hat{p}_t \dots p_n; p_t q_1 \dots q_m) \\ &\quad - \sum_s H(p_1 \dots p_n q_s; q_1 \dots \hat{q}_s \dots q_m) \qquad n \geqq 0, m \geqq 1. \end{aligned} \tag{13}$$

Equation (11) itself has been encountered before: we note again that it is the statement of the Goldstone theorem. As an exercise let us consider the case $n = 1, m = 1$. We shall write $H(p; q_1, q_2)$ in terms of irreducible $\sigma\pi^2$-vertex $\Gamma_{1,2}(p; q_1, q_2)$:

$$\begin{aligned} &H(p; q_1, q_2) = i\Delta_\sigma(p^2)\, i\Delta_\pi(q_1^2)\, i\Delta_\pi(q_2^2)\, \Gamma_{1,2}(p; q_1, q_2), \\ &p + q_1 + q_2 = 0, \\ &\qquad \Delta_\sigma(p^2) = H(p, -p;). \end{aligned} \tag{14}$$

Then Eq. (13) gives

$$v\Gamma_{1,2}(p; 0, -p) = [\Delta_\sigma(p^2)]^{-1} - [\Delta_\pi(p^2)]^{-1}. \tag{15}$$

We verify that Eq. (15) is satisfied exactly in the lowest order since

$$\Gamma(p; 0, -p) = -2\lambda^2 v, \qquad [\Delta_\pi(p^2)]^{-1} = p^2 - m_\pi^2$$

and

$$[\Delta_\sigma(p^2)]^{-1} = p^2 - m_\pi^2 - 2\lambda^2 v^2.$$

The physical content of the divergence equation (1) is exhausted by the identities of Eqs. (10) and (11) [Eqs. (8) and (9) are constraints on the Green's functions involving axial vector currents]. The meaning of Eq. (11) was thoroughly discussed in the preceeding section. Equation (10) is of the form of

$$(n + m + 1) \text{ point function} \propto \frac{1}{v}\Sigma[(n + m) \text{ point functions}]. \tag{16}$$

Now, according to the power counting rule developed in Sec. 3c, $\frac{1}{v} \sim \lambda$ and the l-loop approximation for the $n + m$ point function is of the order $\lambda^{2(l-1)+n+m}$. Therefore we see that Eq. (10), being of the form of Eq. (16), is satisfied in each order of perturbation theory. That is to say, if the $(n + m + 1)$ point function is computed in the l-loop approximation (i.e. to order $\lambda^{2(l-1)+n+m+1}$) and the $(n + m)$ point function is computed in the l-loop approximation (i.e. to order $\lambda^{2(l-1)+n+m}$), then Eq. (16) is identically satisfied. This is the rationale for our adopting the power counting rule for the perturbation expansion.

The off-mass-shell T-matrix for the $n\sigma$ (assuming the σ to be stable) and m pion process is obtained from the connected Green's function $H(p_1 \dots p_n; q_1 \dots q_m)$ by "amputating" the propagators of the external lines:

$$H(p_1 \dots p_n; q_1 \dots q_m) = \prod_{i=1}^{n} i\Delta_\sigma(p_i^2) \prod_{j=1}^{m} i\Delta_\pi(q_j^2)\, T(p_1 \dots p_n; q_1 \dots q_m).$$

A corollary of Eq. (13) is that

$$\lim_{q_1 \to 0} T(p_1 \dots p_n; q_1 \dots q_m)\big|_{\substack{p_1{}^2 = p_2{}^2 = \dots = m_\sigma{}^2 \\ p_2{}^2 = p_3{}^2 = \dots = m_\pi{}^2}} = 0 \tag{17}$$

which means that the T-matrix vanishes as one of the pion momenta goes to zero, provided that all other external particles are on the mass shell. This is usually referred to as Adler's self consistency condition.

Since, as we have noted, Eq. (13) is true in the Goldstone limit ($\gamma = 0, m_\pi^2 = 0$), we have also

$$\lim_{q_1 \to 0} T(; q_1 q_2 \dots q_m)\big|_{q^2{}_2 = \dots = 0} = 0, \qquad \varepsilon = 0. \tag{18}$$

Equation (18) asserts the absence of the infrared divergences in the Goldstone mode. Individual Feynman diagrams are infrared divergent in this case, but the divergent parts must cancel in every order of perturbation theory. Furthermore, the amplitude must vanish in the soft pion limit. I leave it to the reader to verify that, in the one loop approximation individual Feynman diagrams for $\pi\pi$ scattering are infrared divergent, but the sum of Feynman diagrams does not have this divergence, and in fact satisfies the limit of Eq. (18), in the Goldstone mode of symmetry.

5b π, σ-irreducible vertices

We may express the off-mass shell T-matrix in terms of the full σ and pion propagators:

$$i\Delta_\sigma(p^2) = \int d^4\, x e^{ip\cdot x}\langle T(s(x)\, s(0))\rangle_0,$$

$$\delta_{ij} i\Delta_\pi(q^2) = \int d^4 x\, e^{iq\cdot x}\langle T(\pi_i(x)\, \pi_j(0))\rangle_0 \tag{1}$$

and the irreducible vertices which can not be split up into two or more parts (i.e. which cannot be made disconnected) by removing a full pion or σ propagator from them. More precisely we write

$$T(p_1 \dots p_n; q_1 \dots q_m) = \Gamma_{n,m}(p_1 \dots p_n; q_1 \dots q_m) + \text{reducible part}, \tag{2}$$

where $\Gamma_{n,m}$ is the irreducible vertex for n σ's and m pions, and the reducible part can be written in terms of irreducible vertices of lower order and the full propagators. Expressed in terms of the full propagators (1) and irreducible vertices the T-matrix has a tree structure, i.e. the T-matrix can be represented by graph without closed loops.

The purpose of this subsection is to express the constraints implied by the divergence equation on the irreducible vertices. We shall define

$$\Gamma_{0,0} = \Gamma_{0,1} = \Gamma_{1,0} = \Gamma_{1,1} = 0$$

and

$$\Gamma_{2,0}(p, -p;) = [\Delta_\sigma(p^2)]^{-1}, \tag{3}$$

$$\Gamma_{0,2}(;q, -q) = [\Delta_\pi(q^2)]^{-1}.$$

The Goldstone theorem is expressed as

$$\gamma = -v\Gamma_{0,2}(;00). \tag{4}$$

We have already encountered $\Gamma_{1,2}$ in the preceeding section: It satisfies the identity

$$v\Gamma_{1,2}(p; 0, -p) = \Gamma_{2,0}(p, -p;) - \Gamma_{0,2}(;p, -p). \tag{5}$$

To proceed further it is necessary to express first the Ward-Takahashi identities in terms of the T-matrix. Equation (13) of the preceeding subsection may be written as, for $n > 0$, $m \geqq 1$,

$$\begin{aligned} &vT(p_1 \dots p_n; 0q_1 \dots q_m) \\ &= \sum_s i\Delta_\sigma(q_s^2)\, [i\Delta_\pi(q_s^2)]^{-1}\, T(p_1 \dots p_n q_s; q_1 \dots \hat{q}_s \dots q_m) \\ &\quad - \sum_t i\Delta_\pi(p_t^2)\, [i\Delta_\sigma(p_t^2)]^{-1}\, T(p_1 \dots \hat{p}_t \dots p_n; q_1 \dots q_m p_t). \end{aligned} \tag{6}$$

Let T_1 be the sum of the subset of the tree graphs belonging to $T(p_1 \dots p_n; 0q_1 \dots q_m)$ in which the zero momentum pion comes off a branch, as shown in the figure

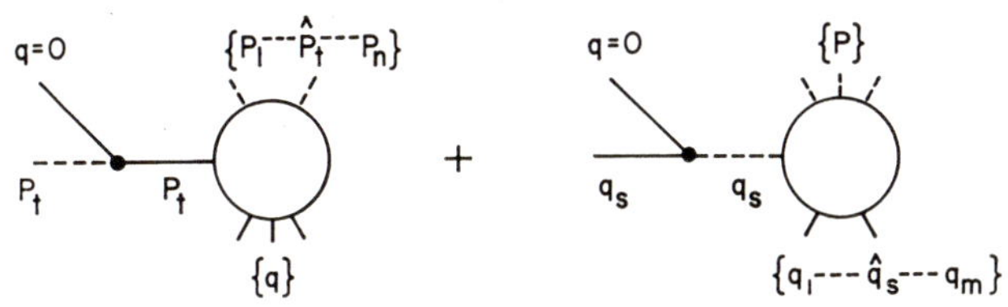

FIGURE 10 Examples of diagrams belonging to T_1: T_1 is the sum of diagrams in which the pion line with $q = 0$ is attached to an external line. Internal lines are full propagators; open circles irreducible vertices.

There are two kinds of graphs belonging to T_1, the typical ones of which are

$$i\Gamma_{1,2}(p_t; 0, -p_t)\, i\Delta_\pi(p_t^2)\, T(p_1 \dots \hat{p}_t \dots p_n; q_1 \dots q_m p_t)$$

and

$$i\Gamma_{1,2}(q_s; 0, -q_s)\, i\Delta_\sigma(q_s^2)\, T(p_1 \dots p_n q_s; q_1 \dots \hat{q}_s \dots q_m).$$

Using the identity (5) we write T_1 as

$$\begin{aligned} vT_1 &= \sum_t T(p_1 \dots \hat{p}_t \dots p_n; q_1 \dots q_m p_t) \\ &\quad - \sum_t [i\Delta_\sigma(p_t^2)]^{-1} [i\Delta_\pi(p_t^2)]\, T(p_1 \dots \hat{p}_t \dots p_n; q_1 \dots q_m p_t) \\ &\quad - \sum_s T(p_1 \dots p_n q_s; q_1 \dots \hat{q}_s \dots q_m) \\ &\quad + \sum_s [i\Delta_\pi(q_s^2)]^{-1} [i\Delta_\sigma(q_s^2)]\, T(p_1 \dots p_n q_s; q_1 \dots \hat{q}_s \dots q_m). \end{aligned} \tag{7}$$

We shall write the complement of T_1 as T_2:

$$T(p_1 \dots p_n; 0q_1 \dots q_m) = T_1 + T_2. \tag{8}$$

When Eqs. (7) and (8) are substituted in Eq. (6), the second and the fourth terms on the right of Eq. (7) cancel the right hand side of Eq. (6), and we obtain

$$\begin{aligned} &vT_2(p_1 \dots p_n; 0q_1 \dots q_m) \\ &\quad = \sum_s T(p_1 \dots p_n q_s; q_1 \dots \hat{q}_s \dots q_m) \\ &\qquad - \sum_t T(p_1 \dots \hat{p}_t \dots p_n; q_1 \dots q_m p_t). \end{aligned} \tag{9}$$

We claim that Eq. (9) is satisfied if the irreducible vertices satisfy

$$
\begin{aligned}
&v\Gamma_{n,m+1}(p_1 \cdots p_n; 0 q_1 \cdots q_m) \\
&\quad = \sum_s \Gamma_{n+1,m-1}(p_1 \cdots p_n q_s; q_1 \cdots \hat{q}_s \cdots q_m) \\
&\qquad - \sum_t \Gamma_{n-1,m+1}(p_1 \cdots \hat{p}_t \cdots p_n; q_1 \cdots q_m p_t), \\
&n \geqq 0, \quad m \geqq 1. \qquad\qquad (10)
\end{aligned}
$$

Equation (10) is true for $n = m = 1$ [see Eq. (5)]. Let us assume that Eq. (10) is true for all n, m; $n + m < N + M$. Consider, now, Eq. (9) for $n = N$, $m = M$. T_2 in Eq. (9) contains two classes of graphs: (a) reducible graphs in which the zero momentum pion comes out of an irreducible vertex contained therein (but not including the graphs in which the zero momentum pion comes out of a three prong irreducible vertex of which two spikes are external lines: such graphs belong to T_1, and not to T_2); (b) irreducible graphs. (See the figure.) For the sum

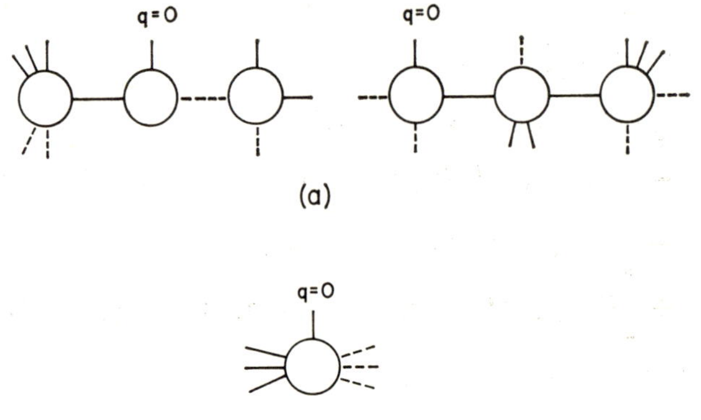

FIGURE 11 Examples of diagrams belonging to T_2. Internal lines are full propagators; open circles irreducible vertices.

of reducible graphs, we may use Eq. (10) for n, m; $n + m < N + M$, to show that the contributions from reducible graphs to both sides of Eq. (9) are identical (the reader should verify this statement at least in simple examples); this leaves only the irreducible vertices on both sides and we obtain Eq. (10) for $n = N$, $m = M$.

It is useful to write Eq. (10) with all the isospin indices restored:

$$v\Gamma^{ii_1\cdots i_m}_{n,m+1}(p_1 \cdots p_n; 0q_1 \cdots q_m)$$
$$= \sum_s \delta^{ii_s}\Gamma^{i_1\cdots i_s\cdots i_m}_{n+1,m-1}(p_1 \cdots p_n q_s; q_1 \cdots \hat{q}_s \cdots q_m)$$
$$- \sum_t \Gamma^{i_1\cdots i_m i}_{n-1,m+1}(p_1 \cdots \hat{p}_t \cdots p_n; q_1 \cdots q_m p_t)$$
$$- \gamma\delta_{n0}\delta_{m,1}\delta^{ii_1}, \tag{11}$$

which is valid for all $n, m \geqq 0$.

We must emphasize at this juncture that the identities (11) are valid in any renormalizable theory in which the divergence equation is valid and the σ and π fields transform like the $[\frac{1}{2}, \frac{1}{2}]$ under the chiral $SU(2) \times SU(2)$ transformation. Whether the nucleon and other fields are included is irrelevant so long as the chiral symmetry is broken in a way to ensure the divergence equation (5a, 1). When there are other fields present, the irreducible vertices we have defined here may still be reducible with respect to these fields.

5c Some simple identities

Equation (11) of the preceeding subsection allows the determination of some irreducible vertices when all external momenta are equal to zero in terms of a few fundamental parameters of the theory.

For this purpose, let us first examine the π and σ propagator. If the pion field is normalized so that

$$\langle 0 \,|\pi^i(x)|\, \pi^j(q)\rangle = \delta^{ij}e^{-iq\cdot x}$$

the pion propagator has the spectral representation

$$\Delta_\pi(p^2) \equiv -i \int d^4x\, e^{ip\cdot x}\langle T(\pi^i(x)\,\pi^i(0))\rangle_0 \qquad (i \text{ not summed})$$

$$= \frac{1}{p^2 - m_\pi^2 + i\varepsilon} + \int_{(3m_\pi)^2}^{\infty} \frac{dm^2}{p^2 - m^2 + i\varepsilon}\, \varrho_\pi(m^2) \tag{1}$$

with a positive definite spectral density $\varrho_\pi(m^2)$, provided that no subtraction is necessary. Then the inverse of the pion propagator behaves in the neighborhood of $p^2 = 0$ as

$$\Delta_\pi^{-1}(p^2) = \Gamma_{0,2}(;p,-p) = -\mu_\pi^2 + \alpha p^2 + 0(p^4), \tag{2}$$

where

$$\frac{1}{\mu_\pi^2} = \frac{1}{m_\pi^2} + \int \frac{dm^2}{m^2} \varrho_\pi(m^2)$$

$$\frac{\alpha}{\mu_\pi^4} = \frac{1}{m_\pi^4} + \int \frac{dm^2}{m^4} \varrho_\pi(m^2). \tag{3}$$

Likewise the σ propagator can be written as

$$\Delta_\sigma(p^2) = -i \int d^4x\, e^{ip\cdot x} \langle T(s(x)\, s(0))\rangle_0$$

$$= \int_{(2m_\pi)^2}^{\infty} \frac{dm^2}{p^2 - m^2 - i\varepsilon} \varrho_\sigma(m^2) \tag{4}$$

with a positive definite spectral density $\varrho_\sigma(m^2)$, assuming that the σ is unstable, and again assuming no subtraction is necessary in writing down Eq. (4). We must have (provided that the quantities exist)

$$1 + \int dm^2\, \varrho_\pi(m^2) = \int dm^2\, \varrho_\sigma(m^2) = \frac{1}{Z_M},$$

since the renormalized fields π and s are proportional to the unrenormalized fields with the same proportionality constant. In the neighborhood of $p^2 = 0$, we shall write

$$\Delta_\sigma^{-1}(p^2) = \Gamma_{2,0}(p,-p;) = -\mu_\sigma^2 + \beta p^2 + 0(p^4), \tag{5}$$

where

$$\frac{1}{\mu^2} = \int \frac{dm^2}{m^2} \varrho_\sigma(m^2); \quad \frac{\beta}{\mu_\sigma^4} = \int \frac{dm^2}{m^4} \varrho_\sigma(m^2). \tag{6}$$

With these preliminaries in mind, let us note first that $\Gamma_{n,m}(p_1 \dots p_n, q_1 \dots q_m) = 0$ unless m is even. Secondly, from the Bose symmetry, we infer that

$$\Gamma_{n,2p} \equiv \Gamma_{n,2p}(\overbrace{0 \dots 0}^{n}; \overbrace{0 \dots 0}^{2p})$$

$$= A_{n,2p}(\delta_{i_1 i_2}\delta_{i_3 i_4} \dots \delta_{i_{2p-1}, i_{2p}} + \cdots), \tag{7}$$

where $A_{n,2p}$ is a constant, and the sum in the bracket is over all possible pairings [(2p − 1)!! ways] of 2p isospin indices. Let us define

$$A_{0,4} = -2\lambda^2 \tag{8}$$

so that the π^4 irreducible vertex can be written as

$$\Gamma_{0,4}^{i_1 i_2 i_3 i_4} = -2\lambda^2(\delta_{i_1 i_2}\delta_{i_3 i_4} + \delta_{i_1 i_3}\delta_{i_2 i_4} + \delta_{i_1 i_4}\delta_{i_2 i_3}). \tag{9}$$

From Eq. (5b, 11) with $n = 0$, $m = 3$

$$v\Gamma_{0,4}^{i_1 i_2 i_3 i_4} = \sum_s \delta^{i_1 i_s}\Gamma_{1,2}^{i_2 \cdots \hat{i}_s \cdots i_4},$$

we find

$$\Gamma_{1,2}^{i_1 i_2} = -2\lambda^2 v\delta_{i_1 i_2}, \tag{10}$$

i.e.

$$A_{1,2} = -2\lambda^2 v, \tag{11}$$

$$\mu_\sigma^2 - \mu_\pi^2 = 2\lambda^2 v^2. \tag{12}$$

We see that with the definition, Eqs. (3), (4) and (8), Eqs. (10) and (12), as well as

$$\gamma = v\mu_\pi^2$$

[which is Eq. (5a, 11)] are *exact* relations which follow from the divergence equation. These relations must *a fortiori* hold in the renormalized σ-model with or without nucleons.

For future reference, it is worth noting that, in general

$$vA_{0,2p+2} = A_{1,2p} \tag{13}$$

as follows from Eq. (5b, 11), a special case of which is Eq. (11).

Bibliography

The use of the Ward-Takahashi identities which follow from the *PCAC* is the main content of the current algebra. Symanzik's method of renormalization consists in determining the subtraction constants of divergent Feynman integrals by a systematic exploitation of the Ward-Takahashi identities.

Our discussion of the derivation of the Ward-Takahashi identities for the irreducible vertices is admittedly sketchy. A more detailed discussion may be found in

1. R. J. Rivers, *J. Math. Phys.* **7**, 385 (1966).

The decomposition of the Feynman amplitudes into irreducible vertices, and conversely, the construction of the former from the latter, are the subject called "structural analysis" by Symanzik. For an introduction to this intricate topic, see for example

2. K. Symanzik's lecture in *Lectures in Theoretical Physics*, W. E. Britten *et al.* (Ed.), (1960).

Adler's self consistency condition was first discussed in

3. S. Adler, *Phys. Rev.* **139**, B 1638 (1965).

In perturbation theory, the Ward-Takahashi identities, Eq. (5a, 3), are ambiguous since both sides of the equation is divergent, and the left hand side of the equation involves a matrix element of an operator which plays no role in the perturbative construction of the T-matrix. To make sense out of this expression in perturbation theory, it is necessary to regularize the expressions. The usual procedure is that of Pauli and Villars, in which unphysical regulator fields are introduced to render integrals finite and unambiguous. See

4. S. N. Gupta, *Proc. Phys. Soc.* **A 63**, 681 (1950); **A 66**, 129 (1952).

Furthermore, for our purpose, the regulator fields must be chosen so that (1) the Ward-Takahashi identities are valid in their presence; (2) the renormalized Green's functions have well-defined and finite limits as the regulator masses tend to infinity. Such a procedure has been found and discussed in

5. J.-L. Gervais and B. W. Lee, *Nuclear Phys.* **B 12**, 627 (1969).

With this kind of regularization, Eq. (5a, 3) is valid and unambiguous. The $k \to 0$ limit may now be taken with impunity to obtain Eq. (5a, 10) which involves only elements of the T-matrix. The regulator masses may now be let to infinity.

A related question is whether the matrix elements of the currents are finite after the renormalization of masses, coupling constants, and wave functions. The answer is yes, but we have not elaborated on this in the text.

VI Construction of soft pion limits—π Irreducible vertices

In this section we shall discuss the construction of the low energy limits of the T-matrix elements for n pions using the machinery prepared in the last section.

We must be precise in what we mean by the low energy limits. We have stated our views that to the zeroth order in ε, which characterizes the breaking of the chiral $SU(2) \times SU(2)$ symmetry, the symmetry is realized in the Goldstone mode, and that the departure from the Goldstone mode, as observed in nature, is small. Equation (5a, 18) tells us that the T-matrix for n pions would vanish in the soft pion limit if the world were exactly in the Goldstone mode. This fact together with the relativistic invariance implies that the T-matrix must vanish as two powers of momentum as all pion momenta go to zero, in the Goldstone limit. Suppose we write $q_i = \xi Q_i$ and let $\xi \to 0$. Then we have

$$\lim_{\xi \to 0} T(; q_1 \dots q_m) = c\xi^2 + 0(\xi^4) \tag{1}$$

in the Goldstone limit. We mean by the low energy limit the expression which is correct to order ξ^2 and ε. We shall ignore terms of order ξ^4, ε^2, $\xi^2\varepsilon$ and higher. Actually, in the limit $\varepsilon = 0$, the expansion in powers of ξ^2 has a zero radius of convergence. However, asymptotically Eq. (1) is still true with the remainder term of the order of $\xi^4(\ln \xi)^p$, $p > 0$. We shall see that the construction of the T-matrix of arbitrary number of pions to this order is unambiguous and self-contained. Actually we shall consider in this section the T-matrix for 2, 4 and 6 pions. A general solution for n pions is postponed until a later section.

We will introduce the concept of π-irreducibility which will facilitate our discussion. A graph is π-irreducible if it cannot be made disconnected by removing a full pion propagator from it. The T matrix for n pions can then be written as

$$T_n(q_1 \dots q_n) \equiv T(; q_1 \dots q_n) = \Pi_n(q_1 \dots q_n) + \pi\text{-reducible part}, \tag{2}$$

where Π_n is the n-th order π-irreducible vertex. We define the second order π-irreducible vertex to be the inverse of the pion propagator

$$\Pi_2(p, -p) = [\Delta_\pi(p^2)]^{-1}$$

or

$$\Pi_2^{ij}(p, -p) = \delta^{ij}[\Delta_\pi(p^2)]^{-1}. \tag{3}$$

The π-irreducible vertex Π_n is in general reducible with respect to the σ-propagator of the last section. Equation (2) implies that the graphs of T_n have tree structures in terms of pion propagators and π-irreducible vertices.

A tree graph for T_n has in general M π-irreducible vertices and $(M - 1)$ pion propagators; as we shall see recursively, all Π_n go to zero like ξ^2 as ξ goes to zero in the Goldstone limit, including Π_2. Therefore, in order to obtain an expression for T_n valid to order ξ^2 and ε, it is necessary and sufficient that Π_n, $n = 2, 4, 6, \dots$ be computed to the same order.

Let us examine $\Pi_2 = \Gamma_{0,2}$. For $\varepsilon = 0$, the pions are Goldstone bosons, and the pion propagator has the form

$$\Delta_\pi(p^2, \varepsilon = 0) = \frac{1}{p^2 + i\varepsilon} + \int\limits_0^\infty \frac{dm^2}{p^2 - m^2 + i\varepsilon} \varrho_\pi(m^2, \varepsilon = 0). \qquad (4)$$

One might wonder if there is an infrared problem in $\Delta_\pi(p^2)$ in the Goldstone limit. Since the G-parity allows the virtual process $\pi \to 3\pi$, 5π etc., but not to 2π, in fact there is no infrared divergence in Eq. (4), and $\Delta_\pi(p^2)$ behaves in the neighborhood of $p^2 = 0$ as

$$\Delta_\pi(p^2, \varepsilon = 0) = \frac{1}{p^2} + \text{const.} + 0(\ln p^2).$$

Therefore, Π_2 is given, in the limit $\varepsilon = 0$, by

$$\Pi_2(\varepsilon = 0) = p^2 + 0(p^4 \ln p^2). \qquad (5)$$

For finite ε, Eq. (5c, 2) gives

$$\Pi_2^{ij} = \delta^{ij}[\alpha p^2 - \mu_\pi^2 + 0(p^4)]. \qquad (6)$$

Comparing Eqs. (5) and (6), and noting that the physical pion mass m_π^2 is the zero of Π_2, we deduce that

$$\alpha = 1 + 0(\varepsilon),$$
$$\mu_\pi^2 = 0(\varepsilon),$$
$$m_\pi^2 - \mu_\pi^2 = 0(\varepsilon^2). \qquad (7)$$

Hence the expression for Π_2^{ij} valid to order ξ^2 and ε, which we shall denote by $\bar{\Pi}_2^{ij}$, is

$$\bar{\Pi}_2^{ij}(p, -p) = \delta^{ij}(p^2 - m_\pi^2). \qquad (8)$$

Note, also, that, since $\gamma = f_\pi m_\pi^2 = v\mu_\pi^2$, we have

$$f_\pi - v = 0(\varepsilon).$$

Let us now consider T_4. Since the G-parity forbids the appearance of a pion propagator in tree graphs, we have

$$T_4 = \Pi_4. \qquad (9)$$

On invariance grounds, the expansion of Π_4 up to and including terms of ξ^2 has the form

$$\begin{aligned}
&\Pi_4^{i_1 i_2 i_3 i_4}(p_1 p_2 p_3 p_4) \\
&\quad = \delta_{i_1 i_2}\delta_{i_3 i_4}[a(p_1 + p_2)^2 + b(p_1 + p_3)^2 + b(p_1 + p_4)^2 + c] \\
&\qquad + \delta_{i_1 i_3}\delta_{i_2 i_4}[a(p_1 + p_3)^2 + b(p_1 + p_2)^2 + b(p_1 + p_4)^2 + c] \\
&\qquad + \delta_{i_1 i_4}\delta_{i_2 i_3}[a(p_1 + p_4)^2 + b(p_1 + p_2)^2 + b(p_1 + p_3)^2 + c]. \qquad (10)
\end{aligned}$$

We have invoked isospin invariance, Bose symmetry, the momentum conservation $p_1 + p_2 + p_3 + p_4 = 0$, and time reversal invariance to deduce this form. To determine the coefficients a, b, and c, we turn to Eq. (5b, 6), which may be written as

$$v\Pi_4^{i_1 i_2 i_3 i_4}(p_1 p_2 p_3 0) = \sum_s \Pi_2^{i_4 i_s}(p_s^2)\, \Delta_\sigma(p_s^2)\, \Gamma_{1,2}^{i_1 \ldots \hat{i}_s \ldots i_3}(p_s; p_1 \ldots \hat{p}_s \ldots p_3). \qquad (11)$$

Since the right hand side of Eq. (11) contains the factor Π_2, it is already of the order of ξ^2 or of ε, and in order to obtain an expression on the right hand side correct up to this order, it is sufficient to compute the remaining factor to ξ^0 and ε^0. We make use of Eq. (5a, 15)

$$v\Gamma_{1,2}(0; 00) = [\Delta_\sigma(0)]^{-1} - [\Delta_\pi(0)]^{-1}$$

to deduce that

$$\Delta_\sigma(0)\, \Gamma_{1,2}(0; 00) = v^{-1} \qquad (\varepsilon = 0) \qquad (12)$$

$\Delta_\pi^{-1}(0)$ being zero in the Goldstone limit. Note that possible infrared divergences in the two factors $\Delta_\sigma(0)$ and $\Gamma_{1,2}$ cancel to give a well-defined constant on the right hand side of Eq. (12). Combining Eqs. (12) and (11), we obtain

$$\begin{aligned}
&\Pi_4^{i_1 i_2 i_3 i_4}(p_1 p_2 p_3 0) \\
&\quad = v^{-2}\{\delta_{i_1 i_2}\delta_{i_3 i_4}(p_3^2 - m_\pi^2) + \delta_{i_1 i_3}\delta_{i_2 i_4}(p_2^2 - m_\pi^2) + \delta_{i_1 i_4}\delta_{i_2 i_3}(p_1^2 - m_\pi^2)\}. \qquad (13)
\end{aligned}$$

Comparing Eqs. (10) and (13), we find $a = 1$, $b = 0$, $c = -m_\pi^2$, so finally

$$\Pi_4^{i_1 i_2 i_3 i_4} = \delta_{i_1 i_2}\delta_{i_3 i_4}\frac{1}{f_\pi^2}[(p_1 + p_2)^2 - m_\pi^2] + \text{Bose permutations.} \qquad (14)$$

Now to the sixth order π-irreducible vertex. T_6 may be written as

$$T_6 = \Pi_6 + \Sigma \Pi_4 \Delta_\pi \Pi_4,$$

where the summation is over all partions of six pions into two groups, each of three. Again, on invariance grounds the expression for Π_6 valid to order ξ^2 and ε is of the form

$$\bar{\Pi}_6 = \Sigma \delta_{12}\delta_{34}\delta_{56}\{a[(p_1 + p_2)(p_3 + p_4) + (p_3 + p_4)(p_5 + p_6) + (p_5 + p_6)(p_1 + p_2)] + b[p_1 p_2 + p_3 p_4 + p_5 p_6] + c\}, \tag{15}$$

where we have used 1, 2, ... for $i_1, i_2, \ldots$ and the summation is over 15 possible pairings of six objects. The analogue of Eq. (11) reads now

$$T_6(p_1 \ldots p_5 0) = \frac{1}{v} \sum_{s=1}^{5} \delta_{6s} \Pi_2(p_s^2) \Delta_\sigma(p_s^2) T(p_s; p_1 \ldots \hat{p}_s \ldots p_5). \tag{16}$$

Again, because Π_2 is of order ξ^2 or ε, the rest of the factors on the right need be evaluated to order ξ_0 and ε^0, in order to compare Eqs. (16) and (15) in the appropriate limit. Furthermore, it will suffice to consider the coefficients of $\delta_{12}\delta_{34}\delta_{56}$ in Eqs. (15) and (16) to compute a, b and c in Eq. (15).

Consider $T(p_5; p_1 \ldots p_4)$. The π-reducibility of this object is such that we can write the coefficient of $\delta_{12}\delta_{34}$ of this quantity which is valid to order ξ^0 and ε^0 as (see figure)

$$\Delta_\sigma(p_5^2) T(p_5; p_1 \ldots p_4) = \delta_{12}\delta_{34}\left\{e - \frac{1}{v^3}\left[\frac{(p_1 + p_2)^2 - m_\pi^2}{(p_1 + p_2 + p_3)^2 - m_\pi^2} + \cdots\right]\right\} + \cdots, \tag{17}$$

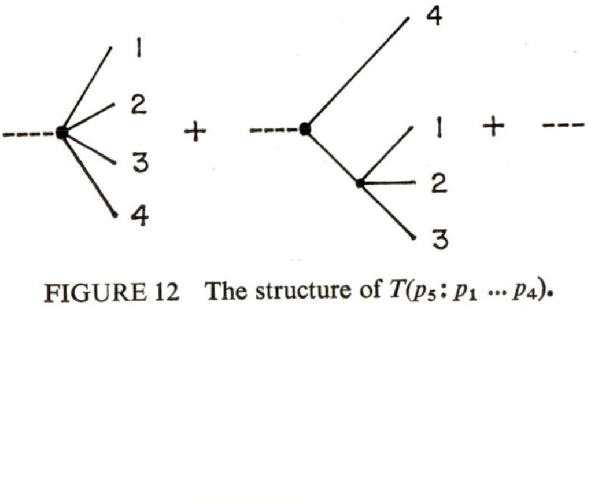

FIGURE 12 The structure of $T(p_5 : p_1 \cdots p_4)$.

where the omitted terms in the square bracket refer to π-reducible terms which may be obtained from the first term by permutations. When Eq. (17) is substituted in Eq. (16), all the terms in the square bracket cancel against π-reducible terms of T_6 on the left. There are, however, two π-reducible terms in T_6 which are not cancelled in this way. These are shown in the figure

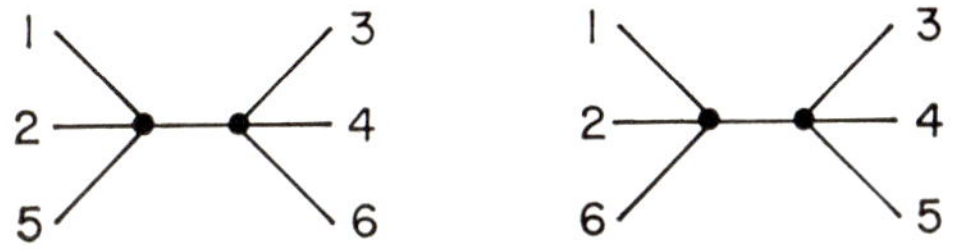

FIGURE 13 The π-reducible diagrams of $T_6(p_1 \dots p_5 \cdot 0)$ which are not cancelled by the right hand side of Eq. (6.16)

and their contribution to T_6 is

$$-\frac{1}{v^4}\left\{\frac{[(p_1+p_2)^2-m_\pi^2][(p_3+p_4)^2-m_\pi^2]}{(p_1+p_2+p_5)^2-m_\pi^2}\right.$$
$$\left.+\frac{[(p_1+p_2)^2-m_\pi^2][(p_3+p_4)^2-m_\pi^2]}{(p_1+p_2+p_6)^2-m_\pi^2}\right\}$$

valid to order ξ^2 and ε. As $p_6 \to 0$, this reduces to

$$-\frac{1}{v^4}[(p_1+p_2)^2-m_\pi^2+(p_3+p_4)^2-m_\pi^2].$$

Therefore, Eq. (16) implies

$$a[(p_1+p_2)\cdot(p_3+p_4)+(p_3+p_4)\cdot p_5+(p_1+p_2)\cdot p_5]+b[p_1\cdot p_2$$
$$+p_3\cdot p_4]+c-\frac{1}{v^4}[(p_1+p_2)^2+(p_3+p_4)^2-2m_\pi^2]=\frac{1}{v}e(p_5^2-m_\pi^2). \tag{18}$$

Since

$$(p_1+p_2)\cdot(p_3+p_4)+p_5\cdot(p_1+p_2+p_3+p_4)$$
$$=-\tfrac{1}{2}p_5^2-\tfrac{1}{2}(p_1+p_2)^2-\tfrac{1}{2}(p_3+p_4)^2,$$

when $p_6 = 0$, Eq. (18) can be solved for a, b, c and e uniquely:

$$a = -\frac{2}{v^4},$$

$$b = 0,$$

$$c = -3\frac{m_\pi^2}{v^4},$$

$$e = -\frac{1}{2}av = \frac{1}{v^3}.$$

Therefore we obtain, at last,

$$\bar{\Pi}_6(p_1 p_2 \dots p_6) = \sum_{\text{pairings}} \delta_{12}\delta_{34}\delta_{56}\left\{-\frac{2}{f_\pi^4}[(p_1+p_2)\cdot(p_3+p_4) + (p_3+p_4)\cdot(p_5+p_6) + (p_5+p_6)\cdot(p_1+p_2)] - 3\frac{m_\pi^2}{f_\pi^4}\right\}. \tag{19}$$

In principle, this method can be applied to construct all of $\bar{\Pi}_n$. The amount of work necessary increases very rapidly as n increases, however. We shall show a more convenient way of evaluating $\bar{\Pi}_n$ later, by constructing a generating functional for $\bar{\Pi}_n$. Let it suffice for the moment to note that the π-irreducible vertices of order 2, 4 and 6 we obtained are the same if we write

$$\mathscr{L} = \frac{1}{2}(\partial\pi)^2 + \frac{1}{2}\frac{(\pi\cdot\partial\pi)^2}{f_\pi^2 - \pi^2} + f_\pi^2\left(\sqrt{f_\pi^2 - \pi^2} - f_\pi\right) \tag{20}$$

and compute the 2, 4 and 6 vertices in the Born approximation, for

$$\mathscr{L}_{(2)} = \frac{1}{2}(\partial\pi)^2 - \frac{m_\pi^2}{2}\pi^2,$$

$$\mathscr{L}_{(4)} = \frac{1}{2f_\pi^2}(\pi\cdot\partial\pi)^2 - \frac{m_\pi^2}{8f_\pi^2}(\pi^2)^2$$

and

$$\mathscr{L}_{(6)} = \frac{1}{2f_\pi^4}(\pi\cdot\partial\pi)^2\,\pi^2 - \frac{m_\pi^2}{16f_\pi^4}(\pi^2)^3.$$

Bibliography

The essential philosophy of this chapter, namely that of expanding the pion amplitudes in powers of ξ^2 and ε, and that of interpreting the *PCAC* as a consequence of a broken Goldstone mode was stressed forcefully in

1. R. Dashen, *Phys. Rev.*, **183**, 1245 (1969).
2. R. Dashen and M. Weinstein, *Phys. Rev.*, **183**, 1261 (1969).

The method we use is somewhat different from that of Dashen and Weinstein. Ours has the virtue of stressing the nonexistence of the infrared catastrophe associated with the zero mass Goldstone bosons; besides, it might be a little bit easier to comprehend than that of Dashen and Weinstein.

The result for the four point function is precisely that of Weinberg,

3. S. Weinberg, *Phys. Rev. Letters*, **17**, 616 (1966)

derived in the context of current algebra.

VII Formalism of Enthalpy Functional—Generating Functional for π, σ-Irreducible Vertices

The first part of this section is devoted to the construction of a generating functional (which we shall call "enthalpy functional") of the irreducible vertices $\Gamma_{n,m}$ of the σ-model. Actually, a generating functional of $\Gamma_{n,m}$ can be constructed very easily from the recursion relation for $\Gamma_{n,m} = \Gamma_{n,m}$ $(0 \ldots 0; 0 \ldots 0)$, Eq. (5b, 11). The presentation here is somewhat roundabout, but is intended to be preliminary to the later construction of a generating functional of π-irreducible vertices $\bar{\Pi}_n$.

Consider a system described by the symmetric version of the σ-model ($\varepsilon = 0$, and to avoid any ambiguities, symmetric in the Wigner mode, i.e., $\mu^2 > 0$) in which however the σ and $\boldsymbol{\pi}$ fields are coupled to c-number external sources $J_\sigma(x)$ and $\mathbf{J}_\pi(x)$. It is sometimes more convenient to adopt the $R(4)$ notations: $\phi = (\sigma, \boldsymbol{\pi})$ and $J = (J_\sigma, \mathbf{J}_\pi)$; $\phi \cdot J = \sigma J_\sigma + \boldsymbol{\pi} \cdot \mathbf{J}_\pi$.

We define the action S by the formula:

$$S = -i \ln \left\langle T \exp \left(i \int d^4x [J(x) \cdot \phi^0(x)] \right) \right\rangle, \tag{1}$$

where $\phi^0(x)$ is the field operator of the symmetric σ-model ($\varepsilon = 0$) in the Heisenberg picture, and $\langle\rangle$ denotes the vacuum expectation value. $\phi^0(x)$ is renormalized as

$$\phi_i^0(x) = Z_M^{-\frac{1}{2}} [\phi_i^0(x)]_{\text{unrenormalized}},$$

where Z_M is defined in Sec. 4. The choice of the finite part of Z_M will be specified later. The action S is a functional of the external sources $J(x)$ and a function of μ^2, among others:

$$S = S[J; \mu^2] = S[J]. \tag{2}$$

The utility of S lies in that it is the generating functional of the Green's functions. We see that

$$\frac{\delta S[J]}{\delta J_i(x)} = e^{-iS}\left\langle T\left(\phi_i^0(x)\, e^{i\int J\cdot\phi^0 d^4z}\right)\right\rangle$$

$$\frac{\delta^2 S[J]}{\delta J_i(x)\,\delta J_j(y)} = i\left[e^{-iS}\left\langle T\left(\phi_j^0(x)\,\phi_i^0(y)\, e^{i\int J\cdot\phi^0 d^4z}\right)\right\rangle\right.$$

$$\left. - e^{-2iS}\left\langle T\left(\phi_i^0(x)\, e^{i\int J\cdot\phi^0 d^4z}\right)\right\rangle\left\langle T\left(\phi_j^0(y)\, e^{i\int J\cdot\phi^0 d^4z}\right)\right\rangle\right] \tag{3}$$

etc., so that as $J \to 0$, we have

$$\left.\frac{\delta^n S[J]}{\delta J_i(x)\,\delta J_j(y)\,\ldots}\right|_{J=0} = (i)^{n-1}\langle T(\phi_i^0(x)\,\phi_j^0(y)\,\ldots)\rangle^c, \tag{4}$$

where the superscript c denotes the connected part of the Green's function.

We shall note an invariance property of $S[J]$. Since the Lagrangian is invariant under the chiral $SU(2) \times SU(2) \sim R(4)$:

$$\begin{aligned} \delta\sigma^0 &= -\boldsymbol{\beta}\cdot\boldsymbol{\pi}^0, \\ \delta\vec{\pi}^0 &= -\boldsymbol{\alpha}\times\boldsymbol{\pi}^0 + \boldsymbol{\beta}\sigma^0, \\ \delta\psi^0 &= [i\boldsymbol{\alpha}/2\cdot\boldsymbol{\tau} - i\boldsymbol{\beta}/2\cdot\boldsymbol{\tau}\gamma_5]\,\psi^0, \end{aligned} \tag{5}$$

and so is the source term $J\cdot\phi^0$, if we endow J with the transformation laws

$$\begin{aligned} \delta J_\sigma &= -\boldsymbol{\beta}\cdot\mathbf{J}_\pi, \\ \delta\mathbf{J}_\pi &= -\boldsymbol{\alpha}\times\mathbf{J}_\pi + \boldsymbol{\beta}J_\sigma, \end{aligned} \tag{6}$$

the action is invariant under the transformations of Eq. (6):

$$\delta_{\boldsymbol{\alpha}} S[J] = \delta_\beta S[J] = 0. \tag{7}$$

From the definition of the currents [see Eq. (2, 5)] we see that

$$-\frac{\delta S[J]}{\delta\,\partial_\mu\boldsymbol{\beta}(x)} = e^{-iS}\left\langle T\left(\mathbf{A}_\mu(x)\, e^{i\int d^4z\phi^0\cdot J}\right)\right\rangle,$$

by considering a response of the action to a space-time dependent (i.e., local) chiral transformation. A similar expression also holds for $\mathbf{V}_\mu$ when $\boldsymbol{\beta}$ is replaced by $\boldsymbol{\alpha}$. We have therefore

$$-\left.\frac{\delta^{n+1} S[J]}{\delta(\partial_\mu \boldsymbol{\beta}(z)\, \delta J_i(x)\, \delta J_j(y) \ldots}\right|_{\substack{\beta=0\\ J=0}} = i^n \langle T(\mathbf{A}_\mu(z)\phi_i^0(x)\phi_j^0(y)\ldots)\rangle^c. \tag{8}$$

To reproduce the Green's functions of the σ-model ($\varepsilon \neq 0$, and in the normal phase), we note first that

$$\begin{aligned}\left.\frac{\delta S[J]}{\delta J_i(x)}\right|_{J=\gamma} &= e^{-iS[\gamma]}\left\langle T\left(\phi_i^0(x)\, e^{i\int \gamma\cdot\phi^0, d^4z}\right)\right\rangle \\ &= \langle \phi_i(x)\rangle,\end{aligned} \tag{9}$$

where γ is a space-time independent four vector in the chiral isospin space: we can choose the direction of γ to be the zeroth axis in the four dimensional chiral axis so that $\gamma \cdot \phi^0 = \gamma\sigma^0$. ϕ_i is the Heisenberg field of the σ-model:

$$\phi_i = e^{-iS[\gamma]}T\left(\phi_i^0 e^{i\int \gamma\sigma^0 d^4z}\right).$$

[This is the interaction picture expression for ϕ_i with the symmetric theory treated as the unperturbed part of the Hamiltonian. See for instance Bjorken and Drell, Vol. II, Eq. (17, 22). I do not believe Haag's theorem is applicable here, because H_0 already contains quartic interactions and $H_I = \int \gamma\sigma^0\, d^3z$ is so much tamer than H_0.] Further, we assume that Z_M has been so chosen that

$$\langle 0\, |\phi_i(x)|\, \pi_j(p)\rangle = e^{-ip\cdot x}\delta_{ij}, \qquad i = 1, 2, 3.$$

Quite generally we have

$$\left.\frac{\delta^n S[J]}{\delta J_i(x)\, \delta J_j(y) \ldots}\right|_{J=\gamma} = (i)^{n-1}\langle T(\bar{\phi}_i(x)\, \bar{\phi}_j(y) \ldots)\rangle^c, \tag{10}$$

where $\bar{\phi}_i = \phi_i - \langle\phi_i\rangle$. Equation (10) means that the expansion coefficients of $S[J]$ about $J = \gamma$ are the connected Green's functions of the σ-model. Since we have argued that the Green's functions are continuous across $\mu^2 = 0$, so is $S[J; \mu^2]$. We may therefore drop the restriction $\mu^2 > 0$ from now on. Obviously the defining formula (1) refers to the symmetric model in the Wigner mode; however, once $S[J; \mu^2]$ is constructed it can be defined for $\mu^2 < 0$ by continuation. It is the generating functional for the Green's functions of the σ-model, by Eq. (10), be it in the normal or Goldstone phase. We note also that the analog of Eq. (8) holds:

$$-\frac{\delta^{n+1}S[J]}{\delta(\partial_\mu\beta^k(z))\,\delta J_i(x)\,\delta J_j(y)\ldots}\Bigg|_{\substack{J=\gamma\\ \beta^k=0}} = i^n\langle T(A_\mu^k(z)\,\bar{\phi}_i(x)\,\bar{\phi}_j(y)\ldots)\rangle^c. \tag{11}$$

We define the c-field $\Phi_i[x; J] = \Phi_i(x)$ by the formula

$$\frac{\delta S[J]}{\delta J_i(x)} = \Phi_i(x). \tag{12}$$

The vacuum expectation value of the ϕ field is given by

$$v_i = \frac{\delta S[J]}{\delta J_i(x)}\bigg|_{J=\gamma} = \langle\phi_i\rangle = \Phi_i[x; J = \gamma], \tag{13}$$

where v_i is a constant (independent of x, as follows from translational invariance of S in the case $J = \gamma =$ const). The invariance property, Eq. (7), of $S[J]$ implies

$$\int d^4x\,\frac{\delta S[J]}{\delta J_i(x)}\,\frac{\delta J_i(x)}{\delta\beta_k} = 0$$

or

$$\int d^4x\left[-\frac{\delta S}{\delta J_\pi^i(x)}\,J_\sigma(x) + \frac{\delta S}{\delta J_\sigma(x)}\,J_\pi^i(x)\right] = 0.$$

Differentiating the above with respect to $J_k(y)$ and letting $J_\sigma \to \gamma$ and $\mathbf{J}_\pi \to 0$, we find

$$v_0 \equiv \langle\sigma\rangle = -\Delta_\pi(0)\,\gamma,$$

$$\mathbf{v} \equiv \langle\boldsymbol{\pi}\rangle = 0.$$

It will prove convenient to treat $\Phi_i(x)$ as independent variables. We shall accomplish this by a functional Legendre transformation:

$$A[\Phi_i] = S[J] - \int J(x)\cdot\Phi(x)\,d^4x, \tag{14}$$

$$\frac{\delta S[J]}{\delta J_i(x)} = \Phi_i(x), \tag{15}$$

$$\frac{A[\Phi]}{\delta\Phi_i(x)} = -J_i(x). \tag{16}$$

In the thermodynamic analogy, we shall call A the enthalpy functional. Differentiating Eq. (16) with respect to $J_j(y)$, we obtain

$$\sum_k \int d^4z \frac{\delta\Phi_k(z)}{\delta J_j(y)} \frac{\delta^2 A[\Phi]}{\delta\Phi_k(z)\,\delta\Phi_i(x)} = -\delta_{ij}\delta^4(x-y). \tag{17}$$

Now, from Eq. (15), we find that

$$\frac{\delta\Phi_k(z)}{\delta J_j(y)} = \frac{\delta^2 S[J]}{\delta J_j(y)\,\delta J_k(z)}.$$

We shall evaluate the left hand side at $J = \gamma$, $\Phi = v$. We obtain

$$\sum_k \int d^4z \left.\frac{\delta^2 S[J]}{\delta J_j(y)\,\delta J_k(z)}\right|_{J=\gamma} \left.\frac{\delta^2 A[\Phi]}{\delta\Phi_k(z)\,\delta\Phi_i(x)}\right|_{\phi=v} = -\delta_{ij}\delta^4(x-y). \tag{18}$$

Since

$$\begin{aligned}\left.\frac{\delta^2 S[J]}{\delta J_j(y)\,\delta J_k(z)}\right|_{J=\gamma} &= i\langle T(\bar\phi_j(y)\,\bar\phi_k(z))\rangle \\ &= -\tilde\Delta_{jk}(y-z),\end{aligned} \tag{19}$$

$$\tilde\Delta_{jk}(y-z) = \delta_{jk}\begin{cases}\tilde\Delta_\pi(y-z), & j = 1, 2, 3\\[2ex] \tilde\Delta_\sigma(y-z) & j = 0\end{cases} \tag{20}$$

$\tilde\Delta_\pi$, $\tilde\Delta_\sigma$ being the full pion and σ propagators in the configuration space, we deduce that

$$\frac{\delta^2 A[\Phi]}{\delta\Phi_k(z)\,\delta\Phi_i(x)} = \delta_{ik}\begin{cases}\tilde\Gamma_{02}(;zx) & i = 1, 2, 3\\[2ex] \tilde\Gamma_{20}(zx;) & i = 0,\end{cases} \tag{21}$$

where

$$\int d^4z\, e^{ip\cdot z}\tilde\Gamma_{02}(;z\cdot 0) = \Gamma_{02}(;p,-p) = \Delta_\pi^{-1}(p^2)$$

and

$$\int d^4z\, e^{ip\cdot z}\tilde\Gamma_{20}(z\cdot 0;) = \Gamma_{20}(p,-p;) = \Delta_\sigma^{-1}(p^2). \tag{22}$$

Differentiating Eq. (17) with respect to $J_k(z)$ once more, we obtain

$$\begin{aligned}&\left.\frac{\delta^3 S[J]}{\delta J_k(z)\,\delta J_j(y)\,\delta J_i(x)}\right|_{J=\gamma}\\ &= \sum_{l,m,n}\int d^4\xi \int d^4\eta \int d^4\zeta \left.\frac{\delta^2 S[J]}{\delta J_k(z)\,\delta J_n(\zeta)}\right|_{J=\gamma} \left.\frac{\delta^2 S[J]}{\delta J_j(y)\,\delta J_m(\eta)}\right|_{J=\gamma}\\ &\quad\times \left.\frac{\delta^2 S[J]}{\delta J_i(x)\,\delta J_l(\xi)}\right|_{J=\gamma} \left.\frac{\delta^3 A[\Phi]}{\delta\Phi_n(\zeta)\,\delta\Phi_m(\eta)\,\delta\Phi_l(\xi)}\right|_{\phi=v},\end{aligned} \tag{23}$$

which shows that

$$[\delta^3 A[\Phi]/\delta\Phi_k(z)\delta\Phi_j(y)\,\delta\Phi_i(x)]|_{\Phi=\gamma}$$

is the third order irreducible vertex. Carrying on this process, we find

$$\delta^{n+m}A[s,p]/\delta s(x_1)\ldots\delta s(x_n)\,\delta p(y_1)\ldots\delta p(y_m)\Big|_{\substack{s=v\\p=0}}$$
$$\equiv \tilde{\Gamma}_{n,m}(x_1\ldots x_n;y_1\ldots y_m), \tag{24}$$

where $s_0 = \Phi_0$, $p_j = \Phi_j$, $j = 1, 2, 3$; $\tilde{\Gamma}_{n,m}$ is the Fourier transform of the irreducible vertex $\Gamma_{n,m}$:

$$\prod_{i=1}^{n}\int d^4x_i\, e^{ip_i\cdot x_i}\prod_{j=1}^{m}\int d^4y_j\, e^{iq_j\cdot y_j}\tilde{\Gamma}_{n,m}(x_1\ldots x_n;y_1\cdots y_m)$$
$$=(2\pi)^4\,\delta^4(\Sigma p_i+\Sigma q_j)\,\Gamma_{n,m}(p_1\ldots p_n;q_1\ldots q_m). \tag{25}$$

What we have shown is this. The expansion coefficients of $A[\Phi]$ about $\Phi = v$ are the irreducible vertices $\Gamma_{n,m}$ where the chiral four vector v is to be determined from

$$\frac{\delta A[\Phi]}{\delta\Phi}\bigg|_{\Phi=v} = -\gamma. \tag{26}$$

If we define the functional B by

$$B[\Phi;\gamma] = A[\Phi] + \gamma\cdot\int d^4x\Phi(x), \tag{27}$$

then we see that

$$\frac{\delta^{n+m}B[\Phi;\gamma]}{\delta s(x_1)\ldots\delta s(x_n)\,\delta p(y_1)\ldots\delta p(y_m)}\bigg|_{\substack{s=v\\ \mathbf{p}=0}} = \tilde{\Gamma}_{n,m}(x_1\ldots x_n;y_1\ldots y_m) \tag{28}$$

where v is to be determined from

$$\frac{\delta B}{\delta s}\bigg|_{s=v,\,\mathbf{p}=0} = 0. \tag{29}$$

Before proceeding further, let us note an invariance of $A[\Phi]$. As noted before, $S[J]$ is invariant under the chiral transformations of Eq. (6). From the definition of $A[\Phi]$, Eqs. (14), and (16), it follows that $A[\Phi]$ is invariant under the transformations:

$$\delta s = -\boldsymbol{\beta}\cdot\mathbf{p},$$
$$\delta\mathbf{p} = -\boldsymbol{\alpha}\times\mathbf{p}+\boldsymbol{\beta}s. \tag{30}$$

Therefore, in particular,

$$\frac{\delta A[\Phi]}{\delta \beta_i} = 0 = \int d^4x \left[\frac{\delta A}{\delta s(x)} p_i(x) - \frac{\delta A}{\delta p_i(x)} s(x) \right].$$

Differentiating further with respect to $p_j(y)$, and letting $s \to v$, $p \to 0$, we obtain

$$-\gamma - \Delta_\pi^{-1}(0)\, v = 0.$$

Let us imagine that we expand all $\Gamma_{n,m}(p_1 \dots p_n; q_1 \dots q_m)$ in powers of momenta:

$$\Gamma_{n,m}(p_1 \cdots p_n; q_1 \cdots q_m) = \Gamma_{n,m} + \Gamma_{n,m}^{(2)} + \cdots, \tag{31}$$

where $\Gamma_{n,m} = \Gamma_{n,m}(0 \dots 0; 0 \dots 0)$ as before; $\Gamma_{n,m}^{(2)}$ is a monomial in invariants $p_i \cdot p_j, p_i \cdot q_j$ and $q_i \cdot q_j$, and so on. We will similarly expand $\tilde{\Gamma}(x_1 \dots x_n; y_1 \dots y_m)$:

$$\tilde{\Gamma}_{n,m}(x_1 \dots x_n; y_1 \dots y_m) = \tilde{\Gamma}_{n,m}^{(0)} + \tilde{\Gamma}_{n,m}^{(2)} \cdots \tag{32}$$

so that

$$\prod_i \int d^4x_i\, e^{ip_i \cdot x_i} \prod_j \int d^4y_j\, e^{iq_j \cdot y_j}\, \tilde{\Gamma}_{n,m}^{(0)}(x_1 \dots x_n; y_1 \dots y_m)$$
$$= (2\pi)^4\, \delta^4(\Sigma p_i + \Sigma q_j)\, \Gamma_{n,m} \tag{33}$$

etc. $\tilde{\Gamma}_{n,m}^{(0)}$ is a constant times a product of delta functions so that it vanishes everywhere in $(R_4)^{n+m}$ except at $x_1 = \cdots = x_n = y_1 \cdots = y_m$; $\tilde{\Gamma}_{n,m}^{(2)}$ is proportional to two derivatives acting on a product of delta functions, and so on.

If we substitute Eq. (32) into the expansion coefficients of $A[\Phi]$ around $\Phi = v$ and collect the terms coming from $\tilde{\Gamma}_{n,m}^{(0)}$ together and call the sum $A^{(0)}$, call the sum of terms coming from $\tilde{\Gamma}_{n,m}^{(2)}$, $A^{(2)}$ and so on, we will have

$$A[\Phi] = A^{(0)}[\Phi] + A^{(2)}[\Phi] + \cdots, \tag{34}$$

where

$$A^{(0)}[\Phi] = \int d^4x \mathscr{A}^{(0)}[\Phi(x)],$$

$$A^{(2)}[\Phi] = \int d^4x \mathscr{A}^{(2)}[\Phi(x)],$$

etc. $\mathscr{A}^{(0)}$ is in general an infinite series of local products of c-fields Φ_i, $\mathscr{A}^{(2)}$ an infinite series of local products of c-fields Φ_i and two derivatives of c-fields and so on.

When the series (34) is truncated at the second order, say, and substituted in Eq. (27):

$$\text{“}B\text{”} = \int d^4x\{\mathscr{A}^{(2)}[\Phi(x)] + \mathscr{A}^{(0)}[\Phi(x)] + \gamma\Phi(x)\}, \tag{35}$$

we obtain what is usually called the “phenomenological Lagrangian”. The meaning of this c-number Lagrangian is simply that, when irreducible vertices $\Gamma_{n,m}$ are computed from it by the algorithm of Eqs. (28) and (25), it gives the expressions which are *correct* to order ξ^2, in the terminology of the last section. [“B” should not be confused with the so-called “nonlinear phenomenological Lagrangian” to be discussed later.]

Furthermore, the vacuum expectation value of Φ one computes from “B” is exact, since

$$\left.\frac{\delta B}{\delta \Phi}\right|_{\Phi=v} = \left.\frac{\delta\text{“}B\text{”}}{\delta \Phi}\right|_{\Phi=v} = \left.\frac{\delta A^{(0)}[\Phi]}{\delta \Phi}\right|_{\Phi=v} + \gamma = 0. \tag{36}$$

The forms of $A^{(0)}$ and $A^{(2)}$, to say nothing of the form of A, is not known, before solving the theory completely. The invariance of A [and therefore of $A^{(0)}$, $A^{(2)}$ etc.], however, specifies the general structures of $A^{(0)}$ and $A^{(2)}$. Let us consider $A^{(0)}$ first. Since there is only one invariant constructed from s and $\mathbf{p}$:

$$s^2 + \mathbf{p}^2 = R(4) \text{ invariant}$$

$\mathscr{A}^{(0)}$ must be of the form

$$\mathscr{A}^{(0)}(x) = \sum_{n=1}^{\infty} f_n(s^2(x) + \mathbf{p}^2(x))^n, \tag{37}$$

where f_n are constants whose values depend on the details of the theory. The vacuum expectation value of the s field, v, is given by Eq. (36):

$$2v \sum_{n=1}^{\infty} n f_n v^{2(n-1)} = -\gamma. \tag{38}$$

$\Gamma_{n,m}^{(0)}$ can be easily obtained from Eq. (37). We write

$$\Gamma_{n,2p}^{i_1\ldots i_{2p}}(0\ldots 0; 0\ldots 0) = \sum_{(2p-1)!!\ \text{pairings}} \delta_{12}\delta_{34}\ldots\delta_{2p-1,2p}A_{n,2p}.$$

Then

$$A_{n,2p} = \left[\left(\frac{d}{dx}\right)\right]^n\left[\left(\frac{d}{dy}\right)^p \sum_{k=1}^{\infty} f_k(x^2 + 2vx + v^2 + 2y)^k\right]_{x=y=0} \tag{39}$$

so that, for example,

$$A_{0,2} = 2\sum_{n=1}^{\infty} nf_n v^{2(n-1)} = -\mu_\pi^2,$$

which gives, when combined with Eq. (38), $v\mu_\pi^2 = \gamma$. The most general form of $\mathscr{A}^{(2)}$ is

$$\mathscr{A}^{(2)}(x) = \left((\partial s)^2 + (\partial \mathbf{p})^2\right)\sum_{n=0}^{\infty} g_n(s^2 + \mathbf{p}^2)^n + (s\,\partial s + \mathbf{p}\,\partial \mathbf{p})^2 \sum_{n=0}^{\infty} h_n(s^2 + \mathbf{p}^2)^n, \tag{40}$$

so that the parameters α and β of Sec. 5c are

$$\alpha = 2\sum_{n=0}^{\infty} g_n v^{2n}$$

and

$$\beta = 2\sum_{n=0}^{\infty} g_n v^{2n} + 2v^2\sum_{n=0}^{\infty} h_n v^{2n}.$$

Bibliography and notes

The generating functional of Green's functions and the subsequent Legendre transformation was first discussed by

1. J. Schwinger, Proc. Nat. Acad. Sci., **37**, 452 (1951).

The presentation in this chapter of the functional Legendre transformation on the action to obtain the generating functional of irreducible vertices follows very closely that of

2. G. Jona-Lasinio, Nuovo Cimento, **34**, 1790 (1964) who also proves the Goldstone theorem by this technique.

The interested reader may verify the main results of Sec. V by the formalism developed in this Section. The Ward-Takahashi identities for the connected Green's functions, Eq. (5a, 10), follow from the expression

$$\left.\frac{\partial S[J]}{\partial \beta_i}\right|_{\beta_i=0} = 0$$

or

$$\int d^4x \left[J_\sigma(x)\frac{\delta S[J]}{\delta J^i_\pi(x)} - J^i_\pi(x)\frac{\delta S[J]}{\delta J_\sigma(x)}\right] = 0.$$

Likewise, the Ward-Takahashi identities of the π,σ-irreducible vertices, Eqs. (5b, 11), follow from

$$\int d^4x \left[s(x) \frac{\delta A}{\delta p^i(x)} - p^i(x) \frac{\delta A}{\delta s(x)} \right] = 0.$$

VIII Enthalpy Functional for π-Irreducible Vertices — Nonlinear Phenomenological Lagrangian

The generating functional for the π-irreducible vertices Π_n can be constructed in much the same way as for the π,σ-irreducible vertices $\Gamma_{n,m}$ discussed in the last section. The π-irreducible vertices Π_n are much more meaningful quantities than $\Gamma_{n,m}$, because, as we have seen in Sec. 6, the expressions for the former valid up to order ξ^2 and ε are unambiguously obtained in terms of two physically observable parameters m_π^2 and f_π, whereas for the latter, an infinite set of parameters are needed to specify them.

Consider the expression

$$\begin{aligned} S_\pi &= -i \ln \left\langle T\, e^{i\int d^4x \eta(x)\cdot \pi(x)} \right\rangle \\ &= -i \ln \left\langle T\, e^{i\int d^4x [\gamma\sigma^0(x) + \eta\cdot\pi^0(x)]} \right\rangle, \end{aligned} \tag{1}$$

where S_π is a functional of the sources $\boldsymbol{\eta} = \mathbf{J}_\pi$. The fields σ^0 and $\boldsymbol{\pi}^0$ refer to the Heisenberg operators in the symmetric σ-model, whereas $\boldsymbol{\pi}$ refer to the Heisenberg field operators of the σ-model with $\varepsilon c = \gamma$.

We shall recapitulate the results of the last section as applied to the functional $S_\pi[\boldsymbol{\eta}]$:

$$\left. \frac{\delta^n S_\pi[\boldsymbol{\eta}]}{\delta\eta_i(x)\,\delta\eta_j(y)\,\ldots} \right|_{\eta=0} = (i)^{n-1} \langle T(\pi_i(x)\,\pi_j(y)\,\ldots)\rangle^c \tag{2}$$

$$- \left. \frac{\delta^{n+1} S_\pi[\boldsymbol{\eta}]}{\delta[\partial_\mu \beta_k(z)]\,\delta\eta_i(x)\,\delta\eta_j(y)\,\ldots} \right|_{\substack{\beta_k=0\\ \eta=0}} = (i)^n \langle T(A_\mu^k(z)\,\pi_i(x)\,\pi_j(y)\,\ldots)\rangle^c. \tag{3}$$

Upon defining

$$A_\pi[\mathbf{p}] = S_\pi[\boldsymbol{\eta}] - \int d^4x\, \boldsymbol{\eta}(x)\cdot \mathbf{p}(x), \tag{4}$$

we obtain

$$\frac{\delta S_\pi[\boldsymbol{\eta}]}{\delta \eta_i(x)} = p_i(x) \tag{5}$$

$$\frac{\delta A_\pi[\mathbf{p}]}{\delta p_i(x)} = -\eta_i(x) \tag{6}$$

$$\left.\frac{\delta^n A_\pi[\mathbf{p}]}{\delta p_i(x)\, \delta p_j(y) \ldots}\right|_{\mathbf{p}=0} \equiv \tilde{\Pi}_n^{ij\ldots}(xy \ldots), \tag{7}$$

where

$$\prod_{i=1}^{n} \int d^4x_i\, e^{ip_i \cdot x_i}\, \tilde{\Pi}_n^{ij\ldots}\,(x_1 x_2 \ldots) = (2\pi)^4\, \delta^4(\Sigma p_i)\, \Pi_n^{ij\ldots}\,(p_1 p_2 \ldots). \tag{8}$$

We further note that, in analogy with Eq. (3),

$$-\left.\frac{\delta^{n+1} A_\pi[\mathbf{p}]}{\delta[\partial_\mu \beta_k(z)]\, \delta p_i(x)\, \delta p_j(y) \ldots}\right|_{\beta_k=0,\, \mathbf{p}=0} \equiv \tilde{\Pi}_\mu^{k;ij\ldots}(z;xy \ldots) \tag{9}$$

in the π-irreducible part of the connected Green's function defined in Eq. (3). Hence

$$A_\mu^k[\mathbf{p}; x] \equiv -\frac{\delta A_\pi[\mathbf{p}]}{\delta\, \partial_\mu \beta^k(x)}, \tag{10}$$

is the generating functional of the irreducible vertices Π_μ of the axial vector current A_μ^k.

Let us now proceed with the construction of $A_\pi[\mathbf{p}]$. Since $A_\pi[\mathbf{p}]$ lacks the chiral invariance, unlike $A[\Phi]$, we must proceed in a somewhat circuitous way: we shall first establish the mapping from $A[\Phi]$ to $A_\pi[\mathbf{p}]$. Note first that

$$S[J] \rightarrow S_\pi[\boldsymbol{\eta}] = S[J]|_{J_\sigma=\gamma,\, \mathbf{J}_\pi=\boldsymbol{\eta}}\,.$$

Therefore, from the defining equations of $A[\Phi]$ and $A_\pi[\mathbf{p}]$, Eqs. (7, 14) and (4), we obtain

$$A[\Phi] \rightarrow A_\pi[\mathbf{p}] = A[\Phi]|_{J_\sigma=\gamma,\, \Phi=\mathbf{p}} + \int \gamma s(x)|_{J_\sigma=\gamma}\, d^4x. \tag{11}$$

The meaning of Eq. (11) is this: $A[\Phi] = A[s, \mathbf{p}]$ is a functional of four independent arguments; $A_\pi[\mathbf{p}]$ is equal to $A[\Phi] + \int d^4x \gamma s(x)$ if we eliminate s by demanding that $J_\sigma(x)$ which is

$$J_\sigma(x) = -\frac{\delta A[s, \mathbf{p}]}{\delta s} \tag{12}$$

is equal to the numerical constant γ:

$$\frac{\delta A[s, \mathbf{p}]}{\delta s(x)} + \gamma = 0. \tag{13}$$

The solution of Eq. (13) is of the form

$$s = s[\mathbf{p}, \gamma]. \tag{14}$$

Thus, we have the desired expression

$$A_\pi[\mathbf{p}] = A[s[\mathbf{p}, \gamma], \mathbf{p}] + \int d^4x \gamma s[\mathbf{p}, \gamma](x), \tag{15}$$

where $s[\mathbf{p}, \gamma](x)$ is to be obtained from Eq. (13). In this way we obtain a *nonlinear* expression of s in terms of $\mathbf{p}$.

The expansion coefficients of $A_\pi[\mathbf{p}]$ around $\mathbf{p} = 0$ are the π-irreducible vertices, $\tilde{\Pi}_n$, whose Fourier transforms are Π_n [see Eqs. (7) and (8)]. If we replace Π_n by $\bar{\Pi}_n$, which are expansions of Π_n to order ξ^2 and ε, in the expansion of $A_\pi[\mathbf{p}]$, we obtain a new functional $\bar{A}_\pi[\mathbf{p}]$, which has the property that

$$\frac{\delta^n \bar{A}_\pi[\mathbf{p}]}{\delta p_i(x)\, \delta p_j(y) \ldots} = \tilde{\bar{\Pi}}_n^{ij\ldots}(xy\ldots), \tag{16}$$

where

$$\prod_{\alpha=1}^{n} \int d^4x_\alpha e^{iq_\alpha . x_\alpha} \tilde{\bar{\Pi}}_n(x_1x_2\ldots) = \bar{\Pi}_n(q_1q_2\ldots)(2\pi)^4\, \delta^4(\Sigma q_\alpha). \tag{17}$$

Since $\tilde{\bar{\Pi}}_n$ contains at most two powers of derivatives, the resulting expression of $\bar{A}_\pi[\mathbf{p}]$ is of the form

$$\bar{A}_\pi[\mathbf{p}] = \int d^4x \Lambda[\mathbf{p}(x)], \tag{18}$$

Λ being in general an infinite series of local products of c-fields, each term containing no more than two derivatives of c-fields. To order ξ^2 and ε, $\bar{A}_\pi[\mathbf{p}]$ agrees with $A_\pi[\mathbf{p}]$. $\bar{A}_\pi[\mathbf{p}]$ is usually referred to as the "nonlinear phenomenological Lagrangian".

In order to obtain $s[\mathbf{p}, \gamma]\,(x)$, we write

$$\frac{\delta}{\delta s(x)}\{A^{(0)}[s, \mathbf{p}] + A^{(2)}[s, \mathbf{p}] + \cdots\} + \gamma = 0, \tag{19}$$

where the general forms of $A^{(0)}$ and $A^{(2)}$ are given in the last section: writing

$$I(x) = s^2(x) + \mathbf{p}^2(x), \tag{20}$$

we have

$$\mathscr{A}^{(0)} = F(I),$$

$$\mathscr{A}^{(2)} = [(\partial s)^2 + (\partial \mathbf{p})^2]\, G(I) + \tfrac{1}{4}(\partial_\mu I)^2\, H(I), \tag{21}$$

where $F(x) = \sum_{n=0} f_n x^n$, etc. We shall write

$$s[\mathbf{p}, \gamma] = s^{(0)}[\mathbf{p}, \gamma] + \xi^2 s^{(2)}[\mathbf{p}, \gamma] + \cdots \tag{22}$$

and substitute it in Eq. (15), and solve it perturbatively in ξ^2 [$A^{(2p)}$ is of order ξ^{2p}]. To the zeroth order in ξ, we have

$$\left.\frac{\delta A^{(0)}}{\delta s(x)}\right|_{s = s^0[\mathbf{p}, \gamma]} = -\gamma \tag{23}$$

or

$$2s(x)\, F'(I) = -\gamma, \qquad F'(x) = \frac{d}{dx} F(x). \tag{24}$$

At this point we recall that

$$\left.\frac{\delta A}{\delta s}\right|_{\substack{s=v \\ \mathbf{p}=0}} = -\gamma \tag{25}$$

is the eigenvalue equation for v. Since

$$\left.\frac{\delta A}{\delta s}\right|_{s=v,\, \mathbf{p}=0} = \left.\frac{\delta A^{(0)}}{\delta s}\right|_{s=v,\, \mathbf{p}=0}, \tag{26}$$

we have

$$2vF'(v^2) = -\gamma \tag{27}$$

[Compare with Eq. (7, 38)]. Since $\gamma = 0(\varepsilon)$, the correct expression for v to order ε^0 is obtained from

$$F'(v^2) = 0, \text{ to order } \varepsilon^0. \tag{28}$$

($v = 0$ would put us in the Wigner mode: remember that in the $\varepsilon = 0$ limit we want the Goldstone mode). Likewise to order ε^0, the constraint Eq. (24) may be written as

$$F'(I) = 0, \text{ to order } \varepsilon^0. \tag{29}$$

Therefore, to order ε^0, Eq. (24) is equivalent to

$$I = v^2$$

or

$$s^0[\mathbf{p}, \gamma](x) = \sqrt{v^2 - \mathbf{p}^2(x)} + 0(\varepsilon), \tag{30}$$

where the square root must be understood as a shorthand notation for its power series expansion.

Let us now write

$$s[\mathbf{p}, \gamma](x) = \sqrt{v^2 - \mathbf{p}^2(x)} + \varepsilon s_1 + \xi^2 s_2 + \cdots \tag{31}$$

combining Eqs. (18) and (26). We claim that, for the purpose of constructing $\bar{A}_\pi[\mathbf{p}]$ to order ε and ξ^2, we do not need to know s_1 and s_2. Returning to Eq. (15), we note that the expression of $A_\pi[\mathbf{p}]$ valid to order ξ^2 and ε are obtained if we write

$$\bar{A}_\pi[\mathbf{p}] \simeq A^{(0)}[s[\mathbf{p}, \gamma], \mathbf{p}] + A^{(2)}[s[\mathbf{p}, \gamma], \mathbf{p}] + \gamma \int s[\mathbf{p}, \gamma](x)\, d^4x \tag{32}$$

or

$$\Lambda[\mathbf{p}(x)] \simeq \mathscr{A}^{(0)}[s[\mathbf{p}, \gamma], \mathbf{p}](x) + \mathscr{A}^{(2)}[s[\mathbf{p}, \gamma], \mathbf{p}](x) + \gamma s[\mathbf{p}, \gamma](x). \tag{33}$$

Since the second and the last term on the right of Eq. (33) are already of order ξ^2 and ε, respectively, it suffices to substitute $s[\mathbf{p}, \gamma] = \sqrt{v^2 - \mathbf{p}(x)^2}$ in them. When Eq. (31) is substituted in $\mathscr{A}^{(0)}$, we have

$$\mathscr{A}^{(0)}[s[\mathbf{p}, \gamma], \mathbf{p}] = F(v^2) + F'(v^2)\, 2\sqrt{v^2 - \mathbf{p}^2}\, (\varepsilon s_1 + \xi^2 s_2) + 0(\varepsilon^2, \varepsilon\xi^2, \xi^4).$$

But, since $F'(v^2) = 0$ to order ε^0 [see Eq. (24)], we see that

$$\mathscr{A}^{(0)}[s[\mathbf{p}, \gamma], \mathbf{p}] = F(v^2) + 0(\varepsilon^2, \varepsilon\xi^2, \xi_4),$$

i.e., the terms εs_1 and $\xi^2 s_2$ do not contribute $\Lambda[\mathbf{p}(x)]$ to order ε and ξ^2.

Since $F(v^2)$, $G(v^2)$ and $H(v^2)$ are constants independent of the c-fields $\mathbf{p}(x)$, and since $\partial_\mu I = 0$, the *unique* expression for $\Lambda[\mathbf{p}(x)]$ is, to within an additive constant,

$$\Lambda[\mathbf{p}(x)] = \tfrac{1}{2}\{(\partial s[\mathbf{p}, 0])^2 + (\partial \mathbf{p})^2\} + f_\pi m_\pi^2 s[\mathbf{p}, 0]$$

where $s[\mathbf{p}, 0] = \sqrt{f_\pi^2 - \mathbf{p}^2(x)}$ [note that $v = f_\pi + 0(\varepsilon)$]. The coefficient of the first term in the above expression is so chosen as to make

$$\Pi_2(p, -p) = p^2 - m_\pi^2.$$

More explicitly, we have

$$\Lambda[\mathbf{p}(x)] = \frac{1}{2}\left\{(\partial\mathbf{p})^2 + \frac{(\mathbf{p}\cdot\partial_\mu\mathbf{p})^2}{f_\pi^2 - \mathbf{p}^2}\right\} + f_\pi m_\pi^2\left[\sqrt{f_\pi^2 - \mathbf{p}^2} - f_\pi\right]. \tag{34}$$

$\bar{A}_\pi[\mathbf{p}]$ constructed from Eq. (34) generates the π-irreducible vertices which are exact to order ξ^2 and ε, in any theory in which

1) the vector and axial vector currents are defined and the charges satisfy the current algebra of chiral $SU(2) \times SU(2)$,
2) the term in the Lagrangian which breaks the chiral symmetry transforms like a component of the $[\frac{1}{2}, \frac{1}{2}]$ representation, and
3) in the limit $\varepsilon = 0$, the symmetry is realized in the Goldstone mode, and the pions are the Goldstone bosons.

The reader may justifiably wonder if the conditions stated above are not too weak. I need only remind him that Eq. (1) could have been written down in any theory satisfying 1) to 3), with the interpretation that $\gamma\sigma$ is the symmetry breaking Lagrangian density, and $\boldsymbol{\pi}$ are the chiral partners of σ which we may use as interpolating fields of pions. Save for some questions of the existence of quantities we wrote down, the analyses of this and the last sections would have gone through just the same. It is remarkable that the π-irreducible vertices depend only on two parameters, to order ξ^2 and ε, and not to any others which may differ from one theory to another.

Finally the generating functional of the irreducible vertices of the axial vector current is obtained from Eq. (10). To order ξ^2 and ε, it is

$$\mathscr{A}^i_\mu[\mathbf{p}](x) = -\{s[\mathbf{p}]\,\partial_\mu p^i - p^i\,\partial_\mu s[\mathbf{p}]\}(x), \tag{35}$$

so that effectively the c-fields $\mathbf{p}(x)$ transform nonlinearly under chiral transformations:

$$\delta\mathbf{p}(x) = \beta\sqrt{f_\pi^2 - \mathbf{p}^2(x)} - \boldsymbol{\alpha}\times\mathbf{p} \tag{36}$$

and

$$\delta\sqrt{f_\pi^2 - \mathbf{p}^2} = -\boldsymbol{\beta}\cdot\boldsymbol{\pi},$$

i.e., the $[\frac{1}{2}, \frac{1}{2}]$ representation is realized by the pion fields $\mathbf{p}$ and their nonlinear function

$$M_{\alpha}^{\dot{\beta}} = \left[\sqrt{f_\pi^2 - \mathbf{p}^2} + i\mathbf{p} \cdot \boldsymbol{\tau}\right]_{\alpha\beta}.$$

The first term on the right hand side of Eq. (34) is invariant under the transformations of Eq. (36). This leads naturally to the subject of "nonlinear realizations of a symmetry group" which we shall not discuss in any further detail, but shall refer to the papers cited in the bibliography.

Bibliography

The interpretation of the nonlinear phenomenological Lagrangian as the generating functional of the π-irreducible vertices, along the line developed in the text, was first proposed by Zumino:

1. B. Zumino, in *Proceedings of Trieste Conference on Renormalization* (1969, unpublished); *Lecture at Seminar on Electromagnetic Interactions and Vector Meson Dominance* (Dubna, 1969, unpublished); *Lectures at Brandeis Summer Institute*, **1970** (to be published).

Zumino communicated to me the general ideas involved in the summer of 1969 at CERN, for which I wish to record my gratitude. The present discussion goes slightly beyond Zumino's original discussion in one aspect: we establish that, in order to obtain the π-irreducible vertices correctly to order ξ^2 and ε, the nonlinear expression for the classical σ-field in terms of the classical π-fields need be known only to the zeroth order in ξ^2 and ε.

In the paper

2. R. Dashen and M. Weinstein, *Phys. Rev.* **183**, 1261 (1969)

he so-called phenomenological Lagrangian is constructed as the generating functional or π-irreducible vertices by a different and very ingenious way.

The nonlinear realization of the chiral symmetry has been discussed by various authors in connection with the phenomenological, or effective, Lagrangian which produces the prediction of current algebra, when used in conjunction with the tree approximation rule:

3. S. Weinberg, *Phys. Rev. Letters*, **18**, 188 (1967).
4. J. Schwinger, *Phys. Letters*, **24 B**, 473 (1967).
5. L. S. Brown, *Phys. Rev.*, **163**, 1802 (1967).
6. W. A. Bardeen and B. W. Lee, *Nuclear and Particle Physics* edited by B. Margolis and C. S. Lam, Gordon and Breach, New York (1967).
7. J. Schwinger, *Phys. Rev.*, **167**, 1432 (1968).

8. J. Wess and B. Zumino, *Phys. Rev.*, **163**, 1727 (1967).
9. S. Weinberg, *Phys. Rev.*, **166**, 1568 (1968).
10. B. W. Lee and H. T. Nieh, *Phys. Rev.*, **166**, 1507 (1968).

The present derivation makes clear the precise nature of approximations used to derive the phenomenological Lagrangian, and the reason for the "tree-approximation" rule.

The general group theoretical problem of nonlinear realizations is discussed in

11. S. Coleman, J. Wess and B. Zumino, *Phys. Rev.*, **177**, 2239 (1969).

This subject has been reviewed comprehensively by

12. S. Gasiorowicz and D. Geffin, *Rev. Mod. Phys.*, **41**, 531 (1969).

IX Nature of chiral $SU(2) \times SU(2)$ breaking

9a Transformation properties of chiral $SU(2) \times SU(2)$ breaking

We have noted that the result of the previous section is quite general for a class of theories in which the chiral $SU(2) \times SU(2)$ current algebra holds, and in wich the chiral symmetry is broken by a term in the Lagrangian which transforms like a zeroth component of a chiral four vector, and the symmetry limit is realized in the Goldstone mode. Let us consider a more general case in which the symmetry breaking Lagrangian transforms according to the $[N/2, N/2]$ representation of the chiral group: we want the isospin preserved, and only the $[N/2, N/2]$ representations contain the $T = 0$ representation of the isospin group. These representations are, furthermore, real.

It is more convenient to adopt an $R(4)$ notation. The basis of the $[N/2, N/2]$ representation is written as an irreducible tensor of $R(4)$:

$$t^{(N)}_{abc\ldots};\ a, b, c \ldots = 0, 1, 2, 3, \tag{1}$$

where the tensor $t^{(N)}$ of the rank N is traceless and symmetric. The symmetry breaking Lagrangian is written as

$$\mathscr{L}' = \gamma t^{(N)}_{000\ldots}(x), \tag{2}$$

where $\gamma = 0(\varepsilon)$. We may define the pion fields to be

$$\pi_i(x) \equiv t^{(N)}_{i00\ldots}(x), \qquad i = 1, 2, 3, \tag{3}$$

$$\langle 0\, | t^{(N)}_{i00\ldots}(x) |\, \pi_j(q)\rangle = \delta_{ij} e^{-iq\cdot x}. \tag{4}$$

Then we have

$$[Q_5^i, t_{000\ldots}^{(N)}] = -iNt_{i00\ldots}^{(N)}, \tag{5}$$

$$[Q_5^i, t_{j00\ldots}^{(N)}] = i\delta_{ij}\left(1 + \frac{N-1}{3}\right) t_{000\ldots}^{(N)} - i(N-1)\, t_{ij00\ldots}^{(N)}, \tag{6}$$

so that

$$[Q_5^i, [Q_5^i, t_{000\ldots}^{(N)}]] = N(N+2)\, t_{000\ldots}^{(N)}. \tag{7}$$

In particular,

$$\partial^\mu A_\mu^i(x) = i[Q_5^i, \mathscr{L}'(x)] = N\gamma\pi^i(x), \tag{8}$$

from which we obtain

$$\gamma = \frac{1}{N} f_\pi m_\pi^2. \tag{9}$$

We shall define $S[J]$ as

$$S[J] = -i \ln \langle T \exp i \int d^4x J^{(N)}(x) \cdot [t^{(N)}(x)]^0 \rangle \tag{10}$$

analogously to Eq. (7, 1), where the superscript 0 denotes the Heisenberg field in the symmetric case ($\gamma = 0$) and

$$J^{(N)} \cdot t^{(N)} = \sum_{a \leqq b \leqq c \leqq \ldots} J_{abc\ldots}^{(N)} t_{abc\ldots}^{(N)} \tag{11}$$

and $A[\Phi]$ by

$$A[\Phi] = S[J] - \int d^4x J^{(N)}(x) \cdot \Phi^{(N)}(x), \tag{12}$$

where

$$\Phi_{abc\ldots}^{(N)}(x) = \frac{\delta S[J]}{\delta J_{abc\ldots}^{(N)}(x)} \tag{13}$$

and

$$J_{abc\ldots}^{(N)}(x) = -\frac{\delta A[\Phi]}{\delta \Phi_{abc\ldots}^{(N)}(x)}. \tag{14}$$

We assume that the Green's functions of $t^{(N)}(x)$'s are well defined.

Denoting $(abc\ \ldots)$ collectively by α, we see that

$$v_\alpha \equiv \langle \Phi_\alpha \rangle|_{J=\gamma} = \delta S/\delta J_\alpha|_{J=\gamma} \tag{15}$$

and

$$\gamma_\alpha = -\delta A/\delta \Phi_\alpha|_{\Phi=v}, \tag{16}$$

where γ is written as a multicomponent vector with only the $\alpha = (000 \cdots)$ component nonvanishing. It is not difficult to show that Eqs. (15) and (16) imply that only the component of v parallel to γ is nonvanishing. Equation (16) is covariant under the chiral transformations

$$\delta\gamma_\alpha = (\boldsymbol{\beta} \cdot \mathbf{T})_{\alpha\beta}\gamma_\beta; \qquad \delta\Phi_\alpha = (\boldsymbol{\beta} \cdot \mathbf{T})_{\alpha\beta}\Phi_\beta, \tag{17}$$

where $(T_i)_{\alpha\beta}$ is the $\left[\frac{N}{2}, \frac{N}{2}\right]$ representation of $-iQ_5^i$: it can be taken to be real and antisymmetric. Therefore, we can write

$$(\delta\gamma)_\alpha = -[\Delta^{-1}(0)]_{\alpha\beta}\,(T_i)_{\beta\gamma}v_\gamma.$$

where $\Delta(p^2)$ is the propagator matrix, and, in particular, in the symmetry limit

$$0 = -[\Delta^{-1}(0)]_{\alpha\beta}\,(T_i)_{\beta\gamma}v_\gamma. \tag{18}$$

Equation (18) states that in the Goldstone mode [$\gamma = 0$, $v_\alpha = 0$, except for $\alpha = (000 \ldots)$, for which case $v_\alpha \neq 0$] the pion mass must be zero, since T_i connects the $T = 0$ state only to the $T = 1$ states: no matter how the symmetry breaking Lagrangian transforms, in the Goldstone limit, the pions, and only the pions become zero mass particles, barring accidental cases, for when α refers to a state of $T > 1$, Eq. (18) simply state $0 = 0$.

We may construct the enthalpy functional $A_\pi[\mathbf{p}]$ as before:

$$A_\pi[\mathbf{p}] = \left\{A[\Phi] + \gamma \int d^4x\Phi^{(N)}_{000\ldots}(x)\right\}\Big|_{J=\gamma}. \tag{19}$$

Equation (19) means that on the right,

$$p_i(x) = \Phi^{(N)}_{i00\ldots}(x)$$

is to be substituted and the other components of Φ_α are to be determined by solving

$$J_\alpha = \gamma_\alpha = -\delta A[\Phi]/\delta\Phi_\alpha, \qquad \alpha \neq (i00 \cdots) \tag{20}$$

for Φ_α, $\alpha \neq (i00 \cdots)$, in terms of p_i and γ. As in the preceeding section, we need the solutions of Eq. (20)

$$\Phi_\alpha(x) = \Phi_\alpha[\mathbf{p}; \gamma]\,(x) \qquad \alpha \neq (i00 \cdots)$$

only to order ξ^0 and ε^0 in order to construct the enthalpy functional correctly to order ξ^2 and ε. Thus $\Phi_\alpha[\mathbf{p}; 0]\,(x)$ may be determined from

$$-\delta A[\Phi]/\delta\Phi_\alpha|_{\Phi_\beta = \Phi_\beta[\mathbf{p};0]} = 0, \tag{21}$$

$$\alpha, \beta \neq (i00 \cdots)$$

and substituted for Φ_α on the right hand side of Eq. (20).

Since $A[\Phi]$ is invariant under the $SU(2) \times SU(2)$ transformations acting on Φ_α, it follows that so must $\{A[\Phi)\}|_{J=\gamma}$ under the transformations acting on p_i and $\Phi_\alpha[\mathbf{p}; 0]$, $\alpha \neq (i00 \cdots)$. In particular, $\{A\}|_{J=\gamma}$ is invariant under the chiral transformations

$$\Phi_\alpha[\mathbf{p}; 0] \rightarrow e^{-i\boldsymbol{\beta}\cdot\mathbf{X}}\Phi_\alpha[\mathbf{p}; 0]\, e^{+i\boldsymbol{\beta}\cdot\mathbf{X}},$$

$$p_i \rightarrow e^{-i\boldsymbol{\beta}\cdot\mathbf{X}} p_i\, e^{i\boldsymbol{\beta}\cdot\mathbf{X}},$$

where X_i is the generator Q_5^i acting on the $\left[\frac{N}{2}, \frac{N}{2}\right]$ representation space spanned by p_i and $\Phi_\alpha[\mathbf{p}; 0]$. The commutation relation between X_i and p_j is given by

$$[X_i, p_j] = i\delta^{ij}\,\frac{N+2}{3}\,\Phi^{(N)}_{000\ldots}[\mathbf{p}; 0]$$

$$-i(N-1)\,\Phi^{(N)}_{ijo\ldots}[\mathbf{p}; 0] \tag{22}$$

and

$$[[X^i, X^j], p^k] = i\varepsilon^{ijl} i\varepsilon^{lkm} p^m. \tag{23}$$

Since $\Phi_\alpha[\mathbf{p}; 0]$ is a nonlinear local function of p_i, the covariance of Eq. (22) under isospin transformations implies that Eq. (22) is of the form of

$$[X^i, p^j] = i\delta^{ij}U(\mathbf{p}^2) + ip^ip^jV(\mathbf{p}^2), \tag{24}$$

$$U(\mathbf{p}^2) + \frac{1}{3}\,\mathbf{p}^2V(\mathbf{p}^2) = \frac{N+2}{3}\,\Phi^{(N)}_{000\ldots}[\mathbf{p}; 0], \tag{25}$$

$$\left(p^ip^j - \frac{1}{3}\,\delta^{ij}\mathbf{p}^2\right)V(\mathbf{p}^2) = -(N-1)\,\Phi^{(N)}_{ijo\ldots}[\mathbf{p}; 0]. \tag{26}$$

The Jacobi identity

$$[X^i, [X^j, p^k]] = [X^j, [X^i, p^k]] + [[X^i, X^j,]\, p^k]$$

demands that U and V satisfy

$$UV - 2UU' - 2xVU' = 1 \tag{27}$$

where $U = U(x)$, $V = V(x)$, and $U' = dU(x)/dx$.

It is possible to define new c-fields $q^i(x)$ by means of nonsingular, nonlinear relations,

$$q^i(x) = p^i(x)\,F(\mathbf{p}^2(x)), \tag{28}$$

$$F(0) = 1,$$

[$F(x)$ must be understood by its power series expansion which we assume exists; the condition $F(0) \neq 0$ assures that Eq. (28) is invertible] so that the new fields $q_j(x)$ has a simpler commutation relation with X^i:

$$[X^i, q^j(x)] = i\delta^{ij} r(x), \tag{29}$$
$$[X^i, r(x)] = -iq^i(x),$$

where $r(x)$ is another function of $p_i(x)$. Since

$$[X^i, F(\mathbf{p}^2(x))] = 2[X^i, p^j]\, p^j F', \tag{30}$$
$$F'(x) = dF(x)/dx,$$

we find that F must satisfy the differential equation;

$$2[F'/F](x) = -\{V(U + xV)^{-1}\}(x) \tag{31}$$

and that $r(x)$ is given by

$$r(x) = U(\mathbf{p}^2(x))\, F(\mathbf{p}^2(x)). \tag{32}$$

The solution of Eq. (31) with the boundary condition $F(0) = 1$ is

$$F(\mathbf{p}^2) = \exp\left\{-\tfrac{1}{2}\int_0^{\mathbf{p}^2} d\lambda V(\lambda)\,[U(\lambda) + \lambda V(\lambda)]^{-1}\right\}$$
$$= v[U^2(\mathbf{p}^2) + \mathbf{p}^2]^{-\frac{1}{2}} \tag{33}$$

where $v = U(0)$ is a constant. We have made use of Eq. (27) in deriving Eq. (33). A simple calculation shows that

$$r^2(x) + \mathbf{q}^2(x) = v^2$$

or

$$r(x) = \sqrt{v^2 - \mathbf{q}^2(x)}. \tag{34}$$

The unique expression of order ξ^2 which is invariant under the chiral transformations implied by Eq. (29) is

$$[\partial r(x)]^2 + [\partial \mathbf{q}(x)]^2,$$

where $r(x)$ is given by Eq. (34). Since the relation between $p_i(x)$ and $q_j(x)$ is uniquely invertible, it follows that the expression for $\{A\}|_{J=\gamma}$ is uniquely given by, to order ξ^2,

$$\{\bar{A}[\Phi]\}|_{J=\gamma} = \int \frac{\alpha}{2}\{(\partial r(x))^2 + (\partial \mathbf{q}(x))^2\}\, d^4x, \tag{35}$$

where α is a constant and r and q_i are to be understood as functions of p_i:

$$q_i(x) = p_i(x)\, v[U^2(\mathbf{p}^2(x)) + \mathbf{p}^2]^{-\frac{1}{2}},$$

$$r(x) = \sqrt{v^2 - \mathbf{q}^2(x)}. \tag{36}$$

It remains to construct the second term inside the curly bracket on the right hand side of Eq. (19). To order ξ^2 and ε we may replace it by

$$\gamma\Phi^{(N)}_{000\ldots}(x) \to \gamma\Phi^{(N)}_{000\ldots}[p; 0]\,(x) \equiv \gamma G(\mathbf{q}^2(x)). \tag{37}$$

Since [see Eqs. (7) and (22)]

$$[X_i, [X_i, \Phi^{(N)}_{000\ldots}[\mathbf{p}; 0]]] = N(N+2)\,\Phi^{(N)}_{000\ldots}[\mathbf{p}; 0], \tag{38}$$

we have the differential equation for G:

$$4(v^2 - x)\, xG''(x) + (6v^2 - 8x)\, G'(x) + N(N+2)\, G(x) = 0. \tag{39}$$

The solution of Eq. (39) may be constructed in powers of $x = \mathbf{q}^2$:

$$G(x) = \text{constant} - \frac{N}{2v}\left[x + \frac{8 - N(N+2)}{20v^2}\, x^2 + \cdots\right],$$

where we used the fact [see Eq. (25)] that

$$G(0) = \frac{3}{N+2}\, U(0) = \frac{3}{N+2}\, v.$$

Now the construction of $\bar{A}_\pi[\mathbf{p}]$ in terms of q_i is complete; by requiring $\bar{\Pi}_2(p, -p) = p^2 - m_\pi^2$ and noting that $\mathbf{q} = \mathbf{p} + 0(p^3)$, and $\gamma = f_\pi m_\pi^2/N$, we find

$$\alpha = 1,$$

$$v = f_\pi.$$

We therefore have

$$\bar{A}_\pi[\mathbf{p}] = \int d^4x \left\{\frac{1}{2}\left[(\partial_\mu \mathbf{q})^2 + \frac{(\mathbf{q}\cdot\partial_\mu \mathbf{q})^2}{f_\pi^2 - \mathbf{q}^2}\right] - \frac{m_\pi^2}{2}\left[\mathbf{q}^2 + \frac{8 - N(N+2)}{20 f_\pi^2}\, (\mathbf{q}^2)^2 + \cdots\right]\right\}. \tag{40}$$

Expressed in terms of $\mathbf{q}$, the above expression differs from the results of the previous section [see Eq. (8, 34]) only in the terms of order ε.

9b Invariance of the *T*-matrix under canonical nonlinear transformations

In order to express $\bar{A}_\pi(\mathbf{p})$ in terms of p_i, we need still to substitute in Eq. (9a, 40) the expression of q_i in terms of p_j, i.e., Eq. (9a, 28). However for the purpose of constructing the *T*-matrix on the mass shell, we may just as well treat the *c*-fields q_i as the pion *c*-fields and obtain π-irreducible vertices from Eq. (9a, 40). We shall elaborate on this remark now. First let us introduce the concept of a *canonical* nonlinear transformation of the *c*-fields. The nonlinear transformation

$$p \to q = X^{-1}(p) \tag{1}$$

is termed canonical if $X^{-1}(p)$ is a power series expansion in p, and $X^{-1}(0) = 0$ $[X^{-1}]'(0) = \{dX^{-1}/dp\}(0) = 1$ [what is really meant is $[X^{-1}]'(0) \neq 0$: then we can always make $[X^{-1}]'(0) = 1$ by a scale transformations on q], i.e.,

$$\delta q(x)/\delta p(y)|_{p=0} = \delta^4(x - y). \tag{2}$$

In such a case, Eq. (1) can be uniquely inverted and p may be expressed as a power series expansion in q:

$$p = X(q); \qquad X(0) = 0, \qquad X'(0) = \frac{dX}{dq}(0) = 1. \tag{3}$$

Now consider the enthalpy functional defined by

$$A[p] = S[\eta] - \int d^4xp\eta.$$

Suppose that we make a canonical nonlinear transformation (3) on p and obtain $A'[q]$ where

$$A[p] = A[X(q)] \equiv A'[q]. \tag{4}$$

We may define the new source ζ by

$$\zeta(x) = -\delta A'[q]/\delta q(x). \tag{5}$$

It is not difficult to see that the nonlinear mapping,

$$\eta = Y[\zeta] \tag{6}$$

which is in general nonlocal, is nevertheless canonical in the sense of

Eq. (2). From Eq. (5) follows the relation

$$\begin{aligned}\zeta(x) &= -\int [\delta A/\delta p(y)]\,[\delta p(y)/\delta q(x)]\,d^4y \\ &= +\int d^4y\eta(y)\,\delta^4(x-y)\,[X(q(x))]' \\ &= \eta(x)\,[X(q(x))]'.\end{aligned} \tag{7}$$

From Eq. (5), we obtain an expression of $q(x)$ in terms of $\zeta(x)$:

$$q(x) = q[\zeta]\,(x). \tag{8}$$

When Eq. (8) is inserted in Eq. (7), we obtain Eq. (6) with the property that

$$[\delta\eta(x)/\delta\zeta(y)]|_{\zeta=0} = \delta^4(x-y) \tag{9}$$

since $X'(0) = 1$ and $\zeta = 0$ implies $q[\zeta] = 0$.

From Eqs. (4) and (5), we can construct $S'[\zeta]$ by the inverse Legendre transformation:

$$S'[\zeta] \equiv A'[q] + \int d^4x\zeta q. \tag{10}$$

The functional differentiation of $S'[\zeta]$ gives the connected T-matrix which is the same as the one constructed from irreducible vertices implied by $A'[q]$ by the usual rules of the tree graph construction.

The statement to be proved is that the T-matrix one obtains from $S'[\zeta]$ by "amputating" the external legs and putting the external momenta on the mass shell is identical to the true T-matrix, i.e., the one obtained from $S[\eta]$. To show this, first note that

$$S'[\zeta] = S[\eta] + \int d^4x\{\zeta(x)\,\delta S'[\zeta]/\delta\zeta(x) - \eta(x)\,\delta S[\eta]/\delta\eta(x)\}. \tag{11}$$

Differentiating Eq. (11) with respect to $\zeta(x)$ we obtain

$$\int d^4y\,\frac{\delta^2 S'[\zeta]}{\delta\zeta(x)\,\delta\zeta(y)}\,\zeta(y) = \int d^4y\int d^4z\eta(y)\,\frac{\delta^2 S[\eta]}{\delta\eta(y)\,\delta\eta(z)}\,\frac{\delta\eta(z)}{\delta\zeta(x)}. \tag{12}$$

Differentiating with respect to $\zeta(y)$ again and putting $\zeta = 0$ [which implies $\eta = 0$; see Eq. (7)], we obtain, using Eq. (9)

$$\left.\frac{\delta^2 S'[\zeta]}{\delta\zeta(x)\,\delta\zeta(y)}\right|_{\zeta=0} = \left.\frac{\delta^2 S[\eta]}{\delta\eta(x)\,\delta\eta(y)}\right|_{\eta=0}. \tag{13}$$

By repeated differentiations of Eq. (12), we arrive at the formula

$$\left.\frac{\delta^n S'[\zeta]}{\delta\zeta(x)\,\delta\zeta(y)\,\dots}\right|_{\zeta=0} - \left.\frac{\delta^n S[\eta]}{\delta\eta(x)\,\delta\eta(y)\,\dots}\right|_{\eta=0}$$

$$= \text{terms of the form of} \int d^4z \left.\frac{\delta^2\eta(z)}{\delta\zeta(x)\,\delta\zeta(y)}\right|_{\zeta=0} \left.\frac{\delta^{n-1}S[\eta]}{\delta\eta(z)\,\dots}\right|_{\eta=0} \tag{14}$$

etc.

The Fourier transforms of the two terms on the left are connected Green's functions: they have poles whenever an external momentum p is put on the mass shell, $p^2 = m^2$, while this is not true for the right hand side. This fact, together with Eq. (13), which asserts the identity of the single particle propagators in two cases, implies the equality of the on-shell T-matrix in two cases, which is obtained by multiplying the Green's function by the inverse propagators and putting the external momenta on the mass shell. We shall refer to this result as "the invariance of the T-matrix under a canonical nonlinear transformation of the c-fields".

Bibliography

The present discussion of the nature of the chiral $SU(2) \times SU(2)$ breaking parallels the similar discussion in

1. S. Weinberg, *Phys. Rev.*, **166**, 1568 (1968).

The demonstration of the invariance of the T-matrix under nonlinear canonical transformations of the c-fields we presented is perhaps new, even though the basic dea has been around for some time. See for example

2. Y. Nambu, *Phys. Letters* **26 B**, 626 (1966)

3. D. G. Boulware and L. S. Brown, *Phys. Rev.*, **172**, 1628 (1968)

and references cited therein. Note that our proof refers specifically to the c-fields in terms of which the effective Lagrangian (or the generating functional of irreducible vertices) is written, and not to the q-number fields of the Lagrangian. In general, nonlinear transformations we have in mind are not defined for operator fields.

X Some applications

10a Chiral symmetry breaking parameter

What is the dimensionless parameter ε? This is the parameter that measures how far the σ-model is from the Goldstone mode of the symmetry limit. We have noted that the pion mass is of order ε. But the pion mass is not dimensionless. A reasonable statement is that the pion mass in units of a typical hadronic mass is of order ε. In the σ-model, this would mean that $\varepsilon \sim (m_\pi/m_\sigma)^2$. If we take m_σ to be $\sim$700 Mev where the $I = 0$, s-wave $\pi\pi$ phase shift has a broad peak near 90°, then $\varepsilon \sim (0.2)^2$ which is indeed a small parameter.

10b $\pi\pi$ scattering

We are now in a position to compare the experimental data on low energy $\pi\pi$ scattering with our theoretical considerations based on the chiral symmetry. For $\pi\pi$ scattering

$$\pi(p_1, i) + \pi(p_2, j) \leftrightarrow \pi(p_3, k) + \pi(p_4, l)$$

we define the Mandelstam variables

$$s = (p_1 + p_2)^2, \quad t = (p_3 - p_1)^2, \quad u = (p_1 - p_4)^2,$$
$$s + t + u = \sum p_i^2 .$$

The invariant T-matrix, defined as

$$S = \delta_{fi} + i(2\pi)^4 \, \delta^4(p_1 + p_2 - p_3 - p_4)\left[\prod_{i=1}^{4} \frac{1}{(2\pi)^{3/2}} \left(\frac{1}{2p_{i0}}\right)^{1/2}\right] T \tag{1}$$

may be written as

$$T = \delta_{ij}\delta_{kl}A(s; tu) + \delta_{ik}\delta_{jl}A(t; us) + \delta_{il}\delta_{jk}A(u; st). \tag{2}$$

For the $[\frac{1}{2}, \frac{1}{2}]$ symmetry breaking, the result of section 6 [say, from Eq. (14)] gives

$$A(s; tu) = \frac{1}{f_\pi^2}(s - m_\pi^2) + 0(\xi^4, \varepsilon^2, \xi^2\varepsilon). \tag{3}$$

For the $[N/2, N/2]$ symmetry breaking [see Eq. (9a, 40)], we have

$$A(s; tu) = \frac{1}{f_\pi^2}\left[s - \frac{8 - N(N + 2)}{5} m_\pi^2\right] + 0(\xi^4, \varepsilon^2, \xi^2\varepsilon). \tag{4}$$

The low energy limit of the s-wave phase shift computed from Eq. (3), assuming that it continues to be a good approximation for $s \simeq 4m_\pi^2$, is

$$m_\pi a_I = \lim_{s \to 4m_\pi^2} \left(\frac{s}{s - 4m_\pi^2} \right)^{1/2} e^{i\delta_I} \sin \delta_I$$

$$= \begin{pmatrix} 7 \\ -2 \end{pmatrix} \frac{1}{32\pi} \left(\frac{m_\pi}{f_\pi} \right)^2 \quad \begin{matrix} I = 0 \\ I = 2. \end{matrix} \tag{5}$$

The transformation properties of the symmetry breaking term do manifest in the scattering lengths of $\pi\pi$ scattering as seen from Eq. (4). While the very low energy $\pi\pi$ scattering data are not yet reliable, it has been possible to compare the scattering lengths predicted by Eq. (4) with the measured cross-section for the process $\pi^- + p \to n + \pi^+ + \pi^-$ near threshold. If the $\pi\pi$ cross-section is small as predicted by Eq. (4) (for reasonable N), the peripheral and nonperipheral contributions may be expected to be comparable. The nonperipheral contribution may be estimated by chiral symmetry also. It is independent of N. The peripheral contribution is proportional to $|2a_0 + a_2|$. Olesson and Turner [Phys. *Rev. Letters, 20,* 1127 (1968)] have fitted the low energy pion production data of Batusov et al. [*Soviet J. Nuclear Phys.*, ***1***, 314 (1965)] with $|2a_0 + a_2|$ left arbitrary. Their result indicates that $|2a_0 + a_2| = (0.4 \pm 0.1)\, m_\pi^{-1}$, in satisfactory agreement with the prediction of the [1/2, 1/2] symmetry breaking, $0.34 m_\pi^{-1}$.

Speaking quite generally, the $[\frac{1}{2}, \frac{1}{2}]$ symmetry breaking of the chiral $SU(2) \times SU(2)$ symmetry appears to be compatible with all known evidences. The small, negative $I = 2$ scattering length Eq. (5) predicts seems to be well substantiated by the experiment of Baton, Laurens and Reignier [*Nuclear Phys.*, ***B 3***, 349 (1967)].

Basdevant and Lee have constructed the renormalized perturbation series for $\pi\pi$ scattering in the σ-model up to one loop terms and converted the series in a Padé approximant [*Phys. Rev.*, ***D 2***, 1680 (1970); *Phys. Letters*, ***B 13***, 182 (1969)]. The Padé approximant is known to have in most cases a larger region of convergence than the power series expansion from which it is built; the Padé approximant of a partial wave amplitude is known to be exactly unitary in the elastic region $4m_\pi^2 < s < 16m_\pi^2$. The σ-model, without nucleons, has three parameters

m_π^2, f_π and λ^2 (see Sec. 4) in the renormalized version. The first two are given by experiment and the last one λ^2 is a free parameter. They chose it so that the unstable σ-meson has a mass of approximately 700 Mev. The value of λ^2 used was

$$\lambda^2 \sim 5.63.$$

The amplitudes constructed in this way were found to satisfy crossing symmetry almost exactly. The s-wave phase shifts were calculated between the range of 0 to 1 Bev c.m. energy. We show the results in the figure together with the experimental data of Malmud and Schlein [*Proc. of Argonne Conf.* (1969)] for $T = 0$ and those of Baton *et al.* [*loc. cit.*] for $T = 2$.

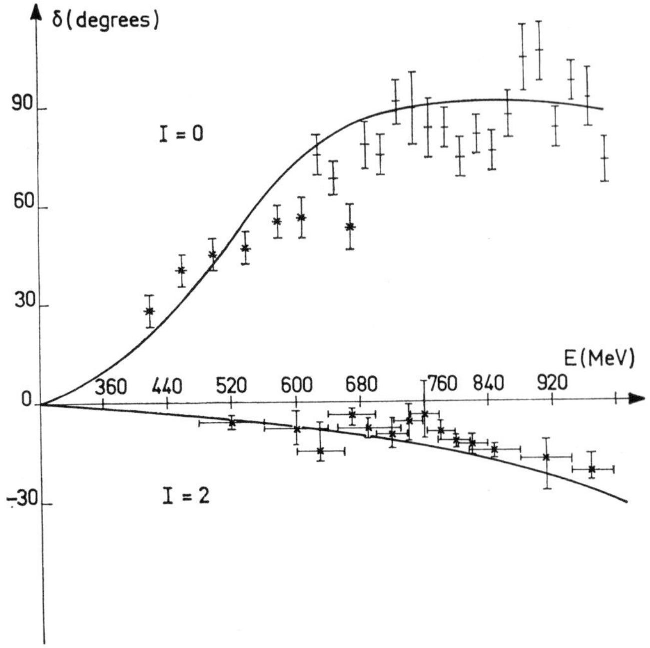

FIGURE 14 The s-wave $\pi\pi$ phase shifts computed from the σ-model by the [1,1] Padé approximant for $f_\pi = 125$ Mev, $\lambda^2 = 5.63$. The upper curve is for $I = 0$. Experimental data are from E. Malamud and P. Schlein, *Proceedings of Argonue Conference*, Argonne National Laboratory (1969), p. 108. The lower one is for $I = 2$. Experimental data are from W. Deinet et al., *Phys. Letters* **30 B**, 359 (1969). The figure is taken from J. L. Basdevant and B. W. Lee, *Phys. Rev.* **D2**, 1680, (1970)

The low energy limit of the s-wave amplitudes may be gleaned from the renormalized Born approximation:

$$\lim_{s\to 4m_\pi^2}\left(\frac{s}{s-4m_\pi^2}\right)^{1/2} e^{i\delta_I}\sin\delta_I$$

$$=\frac{1}{32\pi}\begin{cases} 7\left(\dfrac{m_\pi}{f_\pi}\right)^2\left[1+\dfrac{29}{7}\left(\dfrac{m_\pi}{m_\sigma}\right)^2+\cdots\right] & I=0 \\ -2\left(\dfrac{m_\pi}{f_\pi}\right)^2\left[1-\left(\dfrac{m_\pi}{m_\sigma}\right)^2+\cdots\right] & I=2. \end{cases} \tag{6}$$

As $m_\sigma^2\to\infty$, Eq. (6) reduces to Eq. (5). We expect from Eq. (6) that in the σ-model the s-wave scattering length in the $I = 0(2)$ state is somewhat larger (smaller) in magnitude than that of Eq. (5) [Eq. (5) was first obtained by Weinberg]. This is borne out by the calculation. See the table below

f_π	λ^2	$(a_0)_{B-L}$	$(a_0)_{\text{Weinberg}}$	$(a_2)_{B-L}$	$(a_2)_{\text{Weinberg}}$
95 Mev	7	0.24	0.16	−0.043	−0.046
110 Mev	6.2	0.17	0.12	−0.033	−0.034
125 Mev	5.6	0.12	0.09	−0.025	−0.026

In the table the scattering lengths are expressed in units of m_π^{-1} and they allowed the variations of f_π and λ^2, while keeping the resonance mass of the σ fixed near 700 Mev. The σ scattering amplitudes constructed in this way are almost linear in the interval $0 < s < 4m_\pi^2$, which justifies the ξ^2 approximation in Eq. (3).

The Padé technique applied to higher partial waves in the σ-model produces resonances like ϱ and f_0, for $\lambda^2 \sim 6$, at about the correct energies, but with wrong widths. The approximant we used is of a too low order, for details like widths to be trustworthy. However, the existence of these resonances in the model appears genuine and this may herald an exciting development. Someone should muster enough courage to compute up to three or four loop diagrams; I think this is within the realm of possibility.

A point I should mention before concluding the discussion on $\pi\pi$ scattering: the effective expansion parameter of the perturbation series in the σ-model without nucleons is $\lambda^2/8\pi^2$ which is a small number. This makes the treatment of the model by the Padé technique very reliable. This is not the case with the σ-model including nucleons. There, $g^2/4\pi \sim 15$, so one must proceed to quite higher orders for the calculations to be trustworthy.

10c Inclusion of nucleons—πN Scattering

Let us now include the nucleons in our scheme. We will write the action

$$S[J;\zeta,\bar{\zeta}] = -i\ln\left\langle T\exp\left\{i\int d^4x[J\cdot\phi^{(0)} + \bar{\zeta}\psi^0 + \bar{\psi}^0\zeta]\right\}\right\rangle, \tag{1}$$

where the symbols $\phi^{(0)}$, $\psi^{(0)}$ are defined in Sec. 7. The transformation property of $\psi^{(0)}$ is noted in Eq. (7, 5):

$$\delta\psi^{(0)} = [i\boldsymbol{\alpha}/2\cdot\boldsymbol{\tau} - i\boldsymbol{\beta}/2\cdot\boldsymbol{\tau}\gamma_5]\,\psi^{(0)}. \tag{2}$$

The fermion sources ζ and $\bar{\zeta} = \zeta^+\gamma_0$ are anticommuting c-numbers

$$\{\zeta(x),\zeta(y)\} = \{\bar{\zeta}(x),\zeta(y)\} = \{\bar{\zeta}(x),\bar{\zeta}(y)\} = 0. \tag{3}$$

The nucleon c-field $\Psi(x)$ is defined as

$$\Psi(x) = \frac{\delta}{\delta\bar{\zeta}}\,S[J;\zeta,\bar{\zeta}],$$

$$\bar{\Psi}(x) = \left(\frac{\delta}{\delta\zeta}\right)_R S[J;\zeta,\bar{\zeta}],$$

where the subscipt R denotes that the differentiation should be carried out from right.

We construct the enthalpy functional in the usual manner

$$A[\Phi;\Psi,\bar{\Psi}] = S[J;\zeta,\bar{\zeta}] - \int d^4x[J\cdot\Phi + \bar{\Psi}\zeta + \bar{\zeta}\Psi]. \tag{4}$$

The generating functional $A_{\pi N}[\mathbf{p};\Psi,\bar{\Psi}]$ of the π, N-irreducible vertices is

$$A_{\pi N}[\mathbf{p};\Psi,\bar{\Psi}] = \{A[\Phi;\Psi,\bar{\Psi}] + \gamma\Phi_0\}|_{J=\gamma}, \tag{5}$$

where, inside the curly bracket $\Phi_i = p_i$, $i = 1, 2, 3$ and Φ_0 is to be determined in terms of other variables by the constraint

$$J_0 = \gamma = -\delta A[\Phi;\Psi,\bar{\Psi}]/\delta\Phi_0. \tag{6}$$

We desire to deduce the form of $\bar{A}_{\pi N}$ which is correct to order ε, and to order ξ^2 for pion vertices and to order ξ for pion-nucleon vertices. In momentum space, we shall expand the π,N-irreducible vertices about pion momenta equal to zero as before, and the nucleon momenta $\gamma \cdot p = m$, where m is the physical nucleon mass. We need the solution of Eq. (6), $\Phi_0[p, \Psi, \bar{\Psi}]$, only up to order ε^0 and ξ^0 for this purpose:

$$\{\delta A[\Phi, \Psi, \bar{\Psi}]/\delta\Phi_0\}\,|_{\Phi_0=\Phi_0[p,\Psi,\bar{\Psi}]} = 0.$$

The solution is

$$\Phi_0[\mathbf{p}, \Psi, \bar{\Psi}] = \sqrt{v^2 - \mathbf{p}^2} + f(\bar{\Psi}\Psi), \tag{7}$$

where f is a function of the bilinear invariant $\bar{\Psi}\Psi$. We shall assume that the last term on the right of Eq. (7) may be neglected.

Since $\bar{A}_{\pi N}$ must be invariant under nonlinear chiral transformations:

$$\delta\mathbf{p} = -\boldsymbol{\alpha}\cdot\mathbf{p} + \boldsymbol{\beta}\sqrt{v^2 - p^2}, \qquad \delta\sqrt{v^2 - \mathbf{p}^2} = -\boldsymbol{\beta}\cdot\mathbf{p} \tag{8}$$

and

$$\delta\Psi = \left[\frac{i}{2}\boldsymbol{\alpha}\cdot\boldsymbol{\tau} - \frac{i}{2}\boldsymbol{\beta}\cdot\boldsymbol{\tau}\gamma_5\right]\Psi,$$

the most general form of $\bar{A}_{\pi N}$ which satisfies our specifications is

$$\begin{aligned}\bar{A}_{\pi N}[\mathbf{p}; \Psi, \bar{\Psi}] = \bar{\Psi}\Big[& i\gamma\cdot\partial - g_1(\sigma + i\gamma_5\boldsymbol{\tau}\cdot\mathbf{p}) \\ & + \frac{\alpha}{2f_\pi^2}(\gamma_\mu\boldsymbol{\tau}\cdot\mathbf{p}x\,\partial^\mu\mathbf{p} + \gamma_\mu\gamma_5\boldsymbol{\tau}\cdot(\mathbf{p}\,\partial^\mu\sigma - \sigma\,\partial^\mu\mathbf{p}))\Big]\Psi \\ & + \tfrac{1}{2}[(\partial\sigma)^2 + (\partial\mathbf{p})^2] + f_\pi m_\pi^2(\sigma - f_\pi),\end{aligned} \tag{9}$$

where we have used the abbreviation

$$\sigma = \sqrt{f_\pi^2 - \mathbf{p}^2}. \tag{10}$$

It is important to keep both terms of order ξ^0 and ξ^1 in the pion-nucleon coupling, since

$$\bar{\Psi}\gamma_5\boldsymbol{\tau}\Psi\cdot\mathbf{p} \quad \text{and} \quad i\bar{\Psi}\gamma_\mu\gamma_5\boldsymbol{\tau}\Psi\cdot\partial^\mu\mathbf{p}$$

are both of order ξ^0 when nucleons are on the mass shell. The nucleon mass is $m = g_1 f_\pi$, and the matrix element of the axial vector current between nucleons may be read off from

$$i\frac{\delta \bar{A}_{\pi N}}{\delta\,\partial_\mu \beta^i(x)} = \bar{\Psi}\left[\gamma_5\gamma_\mu \frac{\tau^i}{2}(1+\alpha)\right]\Psi + \cdots$$

i.e.,

$$G_A = 1 + \alpha. \tag{11}$$

The pion-nucleon coupling is given by

$$\bar{\Psi}\left[-ig_1\gamma_5 \boldsymbol{\tau}\cdot\mathbf{p} - i\frac{\alpha}{2f_\pi}\gamma_\mu\gamma_5\boldsymbol{\tau}\,\partial^\mu\mathbf{p}\right]\Psi \approx -ig_1(1+\alpha)\,\bar{\Psi}\gamma_5\boldsymbol{\tau}\Psi\cdot\mathbf{p}. \tag{12}$$

Therefore,

$$g_{\pi NN} = g_1(1+\alpha) + 0(\varepsilon). \tag{13}$$

[There is an error of order ε involved here, because the right hand side is the value of the pion-nucleon vertex at pion momentum $q = 0$, whereas the left hand side is the value at $q^2 = m_\pi^2$]. From Eqs. (11) and (13) follows the Goldberger-Treiman relation

$$g_{\pi NN} f_\pi = G_A m + 0(\varepsilon). \tag{14}$$

The $\pi - N$ scattering is described by the sum of tree graphs: two are nucleon reducible graphs and, in addition, there is a contribution from the irreducible $\bar{N}N\pi^2$ vertex. The low energy limit of the T-matrix is given by

$$S = \delta_{fi} + (2\pi)^4\, i\delta^4(p+q-p'-q')\frac{1}{(2\pi)^6}\sqrt{\frac{m^2}{p_0 p_0'}}\,\frac{1}{\sqrt{4q_0 q_0'}}\,T$$

$$\lim_{q,q'\to 0} T = i\left(\frac{g_{\pi NN}}{2mG_A}\right)^2 \bar{u}(p')\,\gamma\cdot(q+q')\,[\tau_j,\tau_i]\,u(p) \tag{15}$$

for the process $\pi(q,i) + N(p) \to \pi(q',\mathrm{j}) + N(p')$. Translated into scattering lengths, Eq. (15) means

$$m_\pi a_I = \begin{pmatrix} 2 \\ -1 \end{pmatrix}\frac{g_{\pi NN^2}}{8\pi}\left(\frac{m_\pi}{m}\right)^2\frac{1}{G_A^2} + 0(\varepsilon^2), \quad I = \begin{pmatrix} \frac{1}{2} \\ \frac{3}{2} \end{pmatrix}. \tag{16}$$

In fact, Eq. (16) is the Adler-Weisberger relation as interpreted by Tomozawa and Weinberg.

Bibliography

The discussion of $\pi\pi$ scattering in the σ-model is based on the work:

1. J. L. Basdevant and B. W. Lee, *Phys. Letters*, **29B**, 437 (1969); *Phys. Rev.* **D2**, 1680 (1970).

The discussion of πN scattering we presented is very similar to those of

2. J. Schwinger, *Phys. Letters*, **24B**, 473 (1967),
3. W. A. Bardeen and B. W. Lee, *Nuclear and Particle Physics*, edited by B. Margolis and C. S. Lam, Gordon and Breach, New York (1967).

The Adler-Weisberger relation, whose success is responsible for the subsequent development of current algebra and chiral dynamics, appears in

4. W. I. Weisberger, *Phys. Rev. Letters*, **14**, 1047 (1965); *Phys. Rev.*, **143**, 1302 (1966).
5. S. A. Adler, *Phys. Rev. Letters*, **14**, 1051 (1965); *Phys. Rev.*, **140**, B 736 (1965).

XI Currents and Vector Meson Dominance — Phenomenological Lagrangian for Vector Mesons

Sometimes it is necessary to know the matrix elements involving currents, in discussing weak and electromagnetic interactions. We need to know, for example, the Green's function

$$\langle T^*(j_\mu^a(x)\, j_\nu^b(y) \ldots \phi(\xi)\, \phi(\eta) \ldots)\rangle_0, \tag{1}$$

where we shall assume $j_\mu^a(x)$'s are currents satisfying the current algebra:

$$[j_0^a(x), j_0^b(y)]\, \delta(x_0 - y_0) = iC_{abc} j_0^c(x)\, \delta^4(x - y),$$

$$[j_0^a(x), j_i^b(y)]\, \delta(x_0 - y_0) = iC_{abc} j_i^c(x)\, \delta^4(x - y) + \text{s.t.} \tag{2}$$

Here C_{abc} is the structure constant of the relevant group. We shall make no assumptions about the Schwinger terms (s.t.).

In Eq. (1), T^* means that the quantities which follow must be time-ordered, and, if need be, certain contact terms, which are nonvanishing only for $x = y = \cdots = \xi = \eta = \cdots$, must be added to make Eq. (1) Lorentz covariant. An example may be in order. For a conserved currents j_μ, we have the spectral representation of the vacuum expectation value of the time-ordered product of two currents

$$i\int\langle T(j_\mu(x)\,j_\nu(0))\rangle_0\,e^{ik\cdot x}\,d^4x=\left[\int dm^2\,\varrho(m^2)\frac{1}{k^2-m^2}\left(g_{\mu\nu}-\frac{k_\mu k_\nu}{m^2}\right)\right.$$
$$\left.+\,g_{\mu 0}g_{\nu 0}\int\frac{dm^2}{m^2}\varrho(m^2)\right] \tag{3}$$

with $\varrho(m^2)\geqq 0$. We have assumed that no subtraction is necessary in writing down Eq. (3). Further we assumed that the vacuum is invariant under the transformations generated by $Q=\int d^3x j_0(\mathbf{x},t)$. Otherwise, there is an extra term

$$-\frac{k_\mu k_\nu}{k^2}f^2$$

which signifies the contribution of the Goldstone boson to the left hand side of Eq. (3). The time ordered product, Eq. (3), is not covariant.

Multiplying Eq. (3) by k^μ we find

$$\int d^4x\,e^{ik\cdot x}\langle[j_0(x),j_\nu(0)]\rangle_0\,\delta(x_0)=-(g_{\nu\lambda}-\eta_\nu\eta_\lambda)\,k^\lambda\int\frac{dm^2}{m^2}\varrho(m^2), \tag{4}$$

where η is a time-like unit vector: $\eta^2=1$. Now, the left hand side of Eq. (4) is proportional to the matrix element of the Schwinger term. In fact

$$\langle[j_0(x),j_\nu(0)]\rangle_0\,\delta(x_0)=i\,\partial^\lambda S_{\lambda\nu}(x), \tag{5}$$

$$S_{\lambda\nu}(x)\equiv(g_{\lambda\nu}-\eta_\nu\eta_\lambda)\int\frac{dm^2}{m^2}\varrho(m^2)\,\delta^4(x).$$

In this case the T^*-product is defined as

$$T^*(j_\mu(x)\,j_\nu(0))=T(j_\mu(x)\,j_\nu(0))+iS_{\mu\nu} \tag{6}$$

so that

$$i\int\langle T^*(j_\mu(x)\,j_\nu(0))\rangle_0\,e^{ik\cdot x}\,d^4x=\int dm^2\,\varrho(m^2)\frac{1}{k^2-m^2}\left(g_{\mu\nu}-\frac{k_\mu k_\nu}{m^2}\right)$$
$$+\,g_{\mu\nu}\int\frac{dm^2}{m^2}\varrho(m^2) \tag{7}$$

and

$$\partial^\mu\langle T^*(j_\mu(x)\,j_\nu(0))\rangle_0=0. \tag{8}$$

In general the T^*-products are defined to be covariant and to satisfy the identity

$$\partial^\mu \langle T^*(j_\mu^a(x) j_\nu^b(y) \ldots \phi^i(\xi) \ldots)\rangle_0 - \langle T^*(\partial^\mu j_\mu^a(x) j_\nu^b(y) \ldots \phi^i(\xi) \ldots)\rangle_0$$

$$= i \sum \delta^4(x-y)\, C_{abc} \langle T^*(j_\nu^c(y) \ldots \phi^i(\xi) \ldots)\rangle_0$$

$$+ i \sum \delta^4(x-\xi)\, T_{aij} \langle T^*(j_\nu^b(y) \ldots \phi^j(\xi) \ldots)\rangle_0, \qquad (9)$$

where T_{aij} is a representation of $-iQ_a$

$$Q_a = \int d^3x j_0^a(x); \qquad [Q_a]_{ij} = iT_{aij}.$$

For more thorough discussions see the references cited in the bibliography. The hierachy of Eq. (9) can be expressed compactly, if we define

$$S[J; J_\mu] \equiv -i \ln \langle T^* \exp i \int d^4x [J^\mu \cdot j_\mu + J \cdot \phi^0](x)\rangle,$$

$$\left.\frac{\delta^{n+m}}{\delta J_\mu^a(x) \ldots \delta J^i(\xi) \ldots} S[J; J_\mu]\right|_{J_\mu = 0,\, J = \gamma}$$

$$= (i)^{n+m-1} \langle T^*(j_\mu^a(x) \ldots \phi^i(\xi) \ldots)\rangle^c. \qquad (10)$$

Equation (9) is written, then, as

$$\left\{\partial^\mu \frac{\delta}{\delta J_\mu^a(x)} + C_{abc} J_\mu^b(x) \frac{\delta}{\delta J_\mu^c(x)} + T_{aij} J^i(x) \frac{\delta}{\delta J^j(x)}\right\} S[J; J_\mu] = 0. \qquad (11)$$

In verifying Eq. (11), it is necessary to bear in mind that

$$[\partial^\mu j_\mu^a(x)]\,|_{J=\gamma} = -\gamma^i T_{aij} \phi_j(x).$$

An important consequence of Eq. (11) is that the action $S[J; J_\mu]$ is invariant under the local gauge transformations of the sources:

$$\delta J_\mu^a(x) = -C_{bac} \Lambda^b(x)\, J_\mu^c(x) + \partial_\mu \Lambda^a(x),$$

$$\delta J^i(x) = +J^j(x)\, T_{aji} \Lambda^a(x)$$

or

$$\delta J_\mu^a(x)/\delta \Lambda^b(y) = [\delta^{ab}\,\partial_\mu + C_{abc} J_\mu^c(x)]\, \delta^4(x-y),$$

$$\delta J^i(x)/\delta \Lambda^a(y) = J^j(x)\, T_{aji} \delta^4(x-y). \qquad (12)$$

It follows from Eqs. (11) and (12) that

$$\frac{\delta}{\delta \Lambda^a(x)} S[J; J_\mu] = \left\{ [\delta_{ac}\, \partial_\mu + C_{abc} J^b_\mu(x)] \frac{\delta}{\delta J^c_\mu(x)} + J^i(x)\, T_{aij} \frac{\delta}{\delta J^j(x)} \right\} \times S[J; J_\mu] = 0. \tag{13}$$

Let us perform the functional Legendre transformation of $S[J; J_\mu]$ in two steps. First define

$$\begin{aligned} A[\Phi; J_\mu] &= S[J; J_\mu] - \int d^4x J(x) \cdot \Phi(x), \\ \Phi^i(x) &= \delta S / \delta J^i(x), \\ J^i(x) &= -\delta A[\Phi; J_\mu] / \delta \Phi^i(x). \end{aligned} \tag{14}$$

Clearly, $A[\Phi; J_\mu]$ is the generating functional of the vertices involving currents, irreducible with respect to ϕ's. Equation (14) states that $A[\Phi; J_\mu]$ is invariant under the local gauge transformations of J_μ as in Eq. (12) and of Φ contragredient to those of J:

$$\delta \Phi^i(x) = -T_{aij} \Lambda^a(x)\, \Phi^j(x)$$

or

$$\delta \Phi^i(x) / \delta \Lambda^a(x) = -T_{aij} \Phi^j(x).$$

That is,

$$\delta A[\Phi; J_\mu] / \delta \Lambda^a(x) = 0$$

whence follows

$$\left\{ [\delta_{ac}\, \partial_\mu + C_{abc} J^b_\mu(x)] \frac{\delta}{\delta J^c_\mu(x)} - T_{aij} \Phi^j(x) \frac{\delta}{\delta \Phi^i(x)} \right\} A[\Phi : J_\mu] = 0. \tag{15}$$

It is sometimes convenient to perform a further Legendre transformation on $A[\Phi; J_\mu]$

$$B[\Phi, \varrho_\mu] = A[\Phi; J_\mu] - \frac{m_\varrho^2}{2} \int d^4x \left(\varrho_\mu - \frac{1}{\lambda} J_\mu \right)^2 (x) \tag{16}$$

expecially when the propagator, Eq. (7), of conserved vector currents is dominated by a resonance or a particle, with ϱ_μ defined by

$$\delta A[\Phi; J_\mu] / \delta J^a_\mu = -\frac{m_\varrho^2}{\lambda} \left(\varrho^a_\mu - \frac{1}{\lambda} J^a_\mu \right)$$

$$\delta B[\Phi, \varrho_\mu] / \delta \varrho^a_\mu = -m_\varrho^2 \left(\varrho^a_\mu - \frac{1}{\lambda} J^a_\mu \right). \tag{17}$$

In that case ϱ_μ may be regarded as the c-field corresponding to such a particle, with the judicious choice of the parameters m_ϱ^2 and λ. That choice will now be detailed. We define m_ϱ^2 and λ so that

$$\int \frac{dm^2}{m^2} \varrho(m^2) = \frac{m_\varrho^2}{\lambda^2} \tag{18}$$

and

$$\operatorname{Re}\left[\left[\int \frac{dm^2}{m_\varrho^2 - m^2} \varrho(m^2)\right]^{-1}\right] = 0, \tag{19}$$

i.e. m_ϱ^2 is the real part of the (complex) resonance energy.

Now differentiating the first of Eq. (17) with respect to ϱ_λ, we obtain

$$\frac{\delta^2 A}{\delta\varrho_\nu^b(y)\,\delta J_\mu^a(x)} = -\frac{m_\varrho^2}{\lambda}\left\{\delta^{ab} g_{\mu\nu}\,\delta^4(x-y) - \frac{1}{\lambda}\frac{\delta J_\mu^a(x)}{\delta\varrho_\nu^b(y)}\right\}$$

or

$$\int d^4y \left\{\frac{\delta^2 A}{\delta J_\lambda^c(y)\,\delta J_\mu^a(x)} - \frac{m_\varrho^2}{\lambda^2} g_{\mu\lambda}\,\delta^{ac}\,\delta^4(x-y)\right\}\frac{\delta J_\lambda^c(y)}{\delta\varrho_\nu^b(z)}$$

$$= -\frac{m_\varrho^2}{\lambda} g_{\mu\nu}\delta^{ab}\,\delta^4(x-z). \tag{20}$$

Similarly differentiating the second of Eq. (17) with respect to ϱ_ν, we obtain

$$\frac{\delta J_\mu^a(y)}{\delta\varrho_\nu^b(z)} = \frac{\lambda}{m_\varrho^2}\frac{\delta^2}{\delta\varrho_\nu^b(z)\,\delta\varrho_\mu^a(y)}\left\{B[\Phi;\varrho_\mu] + \frac{m_\varrho^2}{2}\int d^4x \sum_i [\varrho_\mu^i(x)]^2\right\}. \tag{21}$$

It follows from Eqs. (7), (10) and (14) that

$$\left[\frac{\delta^2 A}{\delta J_\nu^a(y)\,\delta J_\mu^b(x)} - \frac{m_\varrho^2}{\lambda^2} g_{\mu\nu}\,\delta^4(x-y)\delta^{ab}\right]\Bigg|_{J_\mu=0,\,J=\gamma}$$

$$= \int dm^2 \varrho^{ab}(m^2)\,(g_{\mu\nu} + \partial_\mu\,\partial_\nu/m^2)\,\Delta_F(x-y;m^2)$$

$$\equiv \left(\frac{m_\varrho^2}{\lambda}\right)^2 [-\tilde{\Delta}_\varrho]_{\mu\nu}^{ab}\,(x-y) \tag{22}$$

so that Eq. (20) gives

$$\delta J_\mu^a(y)/\delta\varrho_\nu^b(z)|_{\Phi=v,\,\varrho=0} = \frac{\lambda}{m_\varrho^2}\,[\tilde{\Delta}_\varrho^{-1}]_{\mu\nu}^{ab}\,(y-z),$$

$$\int d^4y [\tilde{\Delta}_\sigma^{-1}]_{\mu\nu}^{ab}\,(x-y)\,[\tilde{\Delta}_\varrho]_{\nu\lambda}^{bc}\,(y-z) = \delta^{ac} g_{\mu\lambda}\,\delta^4(x-z). \tag{23}$$

Combining Eqs. (21) and (23) and defining a new quantity

$$B'[\Phi; \varrho_\mu] = B[\Phi; \varrho_\mu] + \frac{m_\varrho^2}{2} \int d^4x \varrho_\mu^2(x), \tag{24}$$

we obtain

$$\frac{\delta^2}{\delta\varrho_\nu^a(y)\, \delta\varrho_\mu^b(x)} B'[\Phi; \varrho_\mu]\bigg|_{\Phi=v,\, \varrho=0} = [\tilde{\Delta}_\varrho^{-1}]_{\mu\nu}^{ab}\,(x-y). \tag{25}$$

In fact the unusual Legendre transformation, Eq. (16), is chosen so that the second derivative of B' is in the conventional form of the inverse vector propagator, Eq. (25).

While the above discussion is quite general, and does not depend on any particular form of the spectral function $\varrho(m^2)$, it is nevertheless instructive to consider the case in which $\varrho(m^2)$ is approximated by a point spectrum. That such an approximation is valid shall be referred to as the doctrine of vector meson dominance. It appears that one can marshal a considerable amount of experimental supports for such a stand in low-energy phenomenology. Equations (18) and (19) require

$$\varrho(m^2) = \left(\frac{m_\varrho^2}{\lambda}\right)^2 \delta(m^2 - m_\varrho^2). \tag{26}$$

We have therefore

$$\begin{aligned} [\tilde{\Delta}_\varrho]_{\mu\nu}\,(x-y) &= -(g_{\mu\nu} + \partial_\mu\, \partial_\nu/m_\varrho^2)\, \Delta_F(x-y; m_\varrho^2), \\ [\tilde{\Delta}_\varrho^{-1}]_{\mu\nu}\,(x) &= [\partial^2 g_{\mu\nu} - \partial_\mu\, \partial_\nu + m_\varrho^2 g_{\mu\nu}]\, \delta^4(x) \end{aligned} \tag{27}$$

so that the left hand side of Eq. (25) is equal to the inverse of the free vector meson propagator of mass m_σ in the vector meson dominance approximation. In general $[\Delta_\sigma]_{\mu\nu}$ is so normalized that

$$\begin{aligned} [\Delta_\varrho]_{\mu\nu}\,(k) &= \int d^4x\, e^{ip\cdot x}[\tilde{\Delta}_\varrho]_{\mu\nu}\,(x) \\ &= -\int \frac{dm^2}{p^2 - m^2 + i\varepsilon}\, \sigma_\varrho(m^2)\,(g_{\mu\nu} - p_\mu p_\nu/m^2) \\ &\int \frac{dm^2}{m^2}\, \sigma_\varrho(m^2) = \frac{1}{m_\varrho^2}. \end{aligned} \tag{28}$$

As in Sec. 7, repeated differentiations of Eq. (20) with respect to ϱ_λ's and Φ's reveal that $B'[\Phi; \varrho_\mu]$ are the generating functional of the Φ, ϱ-irreducible vertices. More precisely,

$$\text{the } \Phi,\varrho\text{-irreducible part of } \langle T(\underbrace{j^a_\mu(x) \cdots}_{n} \underbrace{\phi^i(\xi) \cdots}_{m})\rangle^c$$

$$= \prod^n \left[\int d^4x' \left(-\frac{m_\varrho^2}{\lambda} \right) [i\tilde{\Delta}_\varrho]^{aa'}_{\mu\mu'} (x - x') \right]$$

$$\times \prod^m \left[\int d^4\xi' i\, \tilde{\Delta}^{ii'}_\phi (\xi - \xi') \right] \times i \frac{\delta^{n+m} B'[\Phi, \varrho_\lambda]}{\delta \varrho^{a'}_{\mu'}(x') \cdots \delta \Phi^{i'}(\xi') \cdots}, \tag{29}$$

where

$$\int d^4\eta\, \tilde{\Delta}^{ik}_\phi (\xi - \eta) \{\delta^2 B'/\delta\Phi^k(\eta)\, \delta\Phi^j(\xi')\}|_{\Phi=\varrho=0} = \delta_{ij}\, \delta^4(\xi - \xi'). \tag{30}$$

In general, the expansion coefficients of $B'[\Phi, \varrho_\mu]$ about $\Phi = \langle\phi\rangle_0$ and $\varrho_\mu = \langle j_\mu \rangle_0 \equiv 0$ are the irreducible vertices of Φ's and ϱ's. The Green's functions of the form of Eq. (1) can be computed from the irreducible vertices by the usual rules of the tree graph construction, and by multiplying the amplitude so constructed by $(-m_\sigma^2/\lambda)^n$, where n is the number of currents appearing in Eq. (1).

How does one construct $B'[\Phi, \varrho_\mu]$ in practice? We must now explore the invariance properties of $B[\Phi, \varrho_\mu]$. From the defining Eq. (16) and Eq. (15), we see immediately that $B[\Phi, \varrho_\mu]$ is invariant under the local gauge transformations

$$\delta\Phi^i(x) = -T_{aij}\Lambda^a(x)\, \Phi^j(x),$$

$$\delta\varrho^a_\mu(x) = -C_{bac}\Lambda^b(x)\, \varrho^c_\mu(x) + \frac{1}{\lambda}\, \partial_\mu \Lambda^a(x). \tag{31}$$

since the last term of Eq. (16) is invariant under the simultaneous *additive gauge transformations* of J_μ and ϱ_μ by $\partial_\mu\Lambda$ and $(1/\lambda)\, \partial_\mu\Lambda$, *characteristic of the classical Yang-Mills fields*. Invariance of the B under the local transformations (31) implies the differential equation for B

$$\{[\delta_{ac}\, \partial_\mu + C_{abc}J^b_\mu(x)]\, \delta/\delta J^c_\mu(x) - T_{aij}\Phi^j(x)\, \delta/\delta\Phi^i(x)\}\; B[\Phi; \varrho_\mu] = 0. \tag{32}$$

Suppose one is to construct B keeping terms up to some powers of ξ. Then one may write

$$\bar{B}[\Phi, \varrho_\mu] = \int d^4x \Lambda[\Phi, \varrho_\mu]\, (x). \tag{33}$$

Since $\Lambda[\Phi, \varrho_\mu](x)$ must be invariant under the local transformation (31), the derivative acting of Φ must appear in the combination

$$D_\mu \Phi_i(x) = [\delta_{ij}\,\partial_\mu + \lambda T_{aij}\varrho^a_\mu(x)]\,\Phi_j(x), \tag{34}$$

where D_μ is the so-called covariant derivative, since

$$\delta[D_\mu \Phi_i(x)] = -T_{aij}\Lambda^a(x)\,[D_\mu \Phi_j(x)].$$

Likewise, ϱ^a_μ itself must appear only in the form

$$\varrho^a_{\mu\nu}(x) = \partial_\mu \varrho^a_\nu(x) - \partial_\nu \varrho^a_\mu(x) - \lambda C_{abc}\varrho^b_\mu(x)\,\varrho^c_\nu. \tag{35}$$

Again, $\varrho^a_{\mu\nu}(x)$ is covariant, in the sense that

$$\delta[\varrho^a_{\mu\nu}(x)] = -C_{bac}\Lambda^b(x)\,\varrho^c_{\mu\nu}(x). \tag{36}$$

Let us apply the procedure to the chiral $SU(2) \times SU(2)$: There are triplets of vector and axial vector currents V^i_μ and A^i_μ. The vector currents are conserved, and the corresponding charges generate the invariance of the vacuum: the isospin group. Therefore, the vacuum expectation value of the time-ordered product of two vector currents is of the form of Eq. (7), assuming no subtraction is called for in writing down the spectral representation. If we call the c-fields corresponding to V^i_μ and A^i_μ, ϱ^i_μ and α^i_μ, the doctrine of vector meson dominance states that these c-fields stand for the mesons which saturate the spectral representations for $\langle T^*(V_\mu V_\nu)\rangle$ and $\langle T^*(A_\mu A_\nu)\rangle$. For the case of V_μ, ϱ_μ's are the c-fields corresponding to the ϱ-mesons.

In order to form the $\varrho^a_{\mu\nu}$ of Eq. (35), it is more convenient to deal with the combinations

$$\boldsymbol{\varrho}^\pm_\mu = \boldsymbol{\varrho}_\mu \pm \boldsymbol{\alpha}_\mu \tag{37}$$

corresponding to $Q^\pm_i = [Q_i \pm Q^5_i]/2$. Then

$$\boldsymbol{\varrho}^\pm_{\mu\nu} = \partial_\mu \boldsymbol{\varrho}^\pm_\nu - \partial_\nu \boldsymbol{\varrho}^\pm_\mu - \lambda \boldsymbol{\varrho}^\pm_\mu \times \boldsymbol{\varrho}^\pm_\nu \tag{38}$$

or, writing

$$\boldsymbol{\varrho}^\pm_{\mu\nu} = \boldsymbol{\varrho}_{\mu\nu} \pm \boldsymbol{\alpha}_{\mu\nu}, \tag{39}$$

we have

$$\boldsymbol{\varrho}_{\mu\nu} = \partial_\mu \boldsymbol{\varrho}_\nu - \partial_\nu \boldsymbol{\varrho}_\mu - \lambda \boldsymbol{\varrho}_\mu \times \boldsymbol{\varrho}_\nu - \lambda \boldsymbol{\alpha}_\mu \times \boldsymbol{\alpha}_\nu,$$

$$\boldsymbol{\alpha}_{\mu\nu} = D_\mu \boldsymbol{\alpha}_\nu - D_\nu \boldsymbol{\alpha}_\mu, \tag{40}$$

where

$$D_\mu = (\partial_\mu - \lambda \boldsymbol{\varrho}_\mu x).$$

The covariant derivatives of the σ, π and nucleon c-fields are defined as

$$\mathscr{D}_\mu s = \partial_\mu s - \lambda \boldsymbol{\alpha}_\mu \cdot \mathbf{p}, \tag{41}$$

$$\mathscr{D}_\mu \mathbf{p} = D_\mu \mathbf{p} + \lambda \boldsymbol{\alpha}_\mu s, \tag{42}$$

$$\mathscr{D}_\mu \Psi = \left(\partial_\mu + \frac{i}{2}\lambda \boldsymbol{\varrho}_\mu \cdot \boldsymbol{\tau} - \frac{i}{2}\lambda \boldsymbol{\alpha}_\mu \cdot \boldsymbol{\tau}\gamma_5\right)\Psi. \tag{43}$$

The phenomenological Lagrangian densities for this system are obtained if we substitute, in the previous expressions, Eqs. (41), (42) and (43) for $\partial_\mu s$, $\partial_\mu \mathbf{p}$ and $\partial_\mu \Psi$ and add the Lagrangian of the gauge fields. If we do not worry about terms quartic in $\bar{\Psi}$, Ψ, ϱ_μ and α_μ, the constraint equation for the (dependent) c-field s is again

$$\begin{aligned} s[\mathbf{p}; \gamma] &= \sqrt{v^2 - \mathbf{p}^2} \\ &= \sqrt{f_\pi^2 - \mathbf{p}^2} + 0(\varepsilon). \end{aligned} \tag{44}$$

Thus the phenomenological Lagrangian density for ϱ's, α's and π's is

$$\begin{aligned} \Lambda[\mathbf{p}; \boldsymbol{\varrho}_\mu, \boldsymbol{\alpha}_\mu] = \Lambda[\boldsymbol{\varrho}_\mu, \boldsymbol{\alpha}_\mu] &+ \frac{\beta}{2}\{[\partial_\mu s - \lambda \boldsymbol{\alpha}_\mu \cdot \mathbf{p})^2 + (D_\mu \mathbf{p} + \lambda \boldsymbol{\alpha}_\mu s)^2\} \\ &+ f_\pi m_\pi^2 (s - f_\pi), \end{aligned} \tag{45}$$

where

$$s = \sqrt{f_\pi^2 - \mathbf{p}^2}$$

and

$$\Lambda[\boldsymbol{\varrho}_\mu, \boldsymbol{\alpha}_\mu] = -\frac{\alpha}{4}\{(\varrho_{\mu\nu})^2 + (\alpha_{\mu\nu})^2\} + \frac{m_\varrho^2}{2}[\varrho_\mu^2 + \alpha_\mu^2]. \tag{47}$$

The inverse ϱ-meson propagator implied by Eq. (47) is

$$[\Delta_\varrho^{-1}]_{\mu\nu}(k) = \alpha(k_\mu k_\nu - g_{\mu\nu}k^2) + g_{\mu\nu} m_\varrho^2 + 0(\xi^4), \tag{48}$$

i.e., α is the coefficient of the ξ^2 term: $(k_\mu k_\nu - g_{\mu\nu}k^2)$, in the expansion of the inverse of the ϱ-meson propagator about $k = 0$, when the propagator is normalized according to Eq. (28). The propagator implied by (48) is

$$[\Delta_\varrho]_{\mu\nu}(k) = -\left(g_{\mu\nu} - \frac{\alpha}{m_\varrho^2}k_\mu k_\nu\right)\frac{1}{\alpha}\frac{1}{k^2 - m_\varrho^2/\alpha} + 0(\xi^2). \tag{49}$$

The vector meson dominance doctrine asserts that $\alpha \simeq 1$.

Before discussing the axial vector propagator it pays to look closely at the term

$$\frac{\beta}{2}(D_\mu \mathbf{p} - \lambda f_\pi \boldsymbol{\alpha}_\mu - \lambda \boldsymbol{\alpha}_\mu s')^2 \tag{50}$$

in Eq. (45), where we have used the abbreviation

$$s' = \sqrt{f_\pi^2 - \mathbf{p}^2} - f_\pi = 0(\mathbf{p}^2). \tag{51}$$

The point is that the cross term of the $D_\mu p$ and $-\lambda f_\pi \boldsymbol{\alpha}_\mu$ is also quadratic in c-fields, so that we need to diagonalize the quadratic term in $D_\mu \mathbf{p}$ and $\boldsymbol{\alpha}_\mu$ which appear in Eq. (45):

$$\begin{aligned}\frac{m_\varrho^2}{2}\boldsymbol{\alpha}_\mu^2 + \frac{\beta}{2}(D_\mu \mathbf{p} + \lambda f_\pi \boldsymbol{\alpha}_\mu)^2 &= \frac{1}{2}[m_\varrho^2 + \beta\lambda^2 f_\pi^2] \\ &\times \left[\boldsymbol{\alpha}_\mu + \frac{\beta\lambda f_\pi}{m_\varrho^2 + \beta\lambda^2 f_\pi^2} D_\mu \mathbf{p}\right]^2 \\ &+ \frac{1}{2}\beta\left[1 - \frac{\beta\lambda^2 f_\pi^2}{m_\varrho^2 + \beta\lambda^2 f_\pi^2}\right](D_\mu \mathbf{p})^2.\end{aligned} \tag{52}$$

Thus, in terms of

$$\boldsymbol{\alpha}'_\mu = \boldsymbol{\alpha}_\mu + \frac{\beta\lambda f_\pi^2}{m_\varrho^2 + \beta\lambda^2 f_\pi^2} D_\mu \mathbf{p}$$

and $D_\mu \mathbf{p}$, the quadratic terms of Eq. (45) are diagonal. This means that the α' and π propagators

$$[\tilde{\Delta}_{\alpha'}^{-1}]_{\mu\nu}^{ij}(x - y) = \frac{\delta^2 C}{\delta\alpha_\mu'^i(x)\,\delta\alpha_\nu'^j(y)} \tag{53}$$

and

$$[\tilde{\Delta}_p^{-1}]^{ij}(x - y) = \frac{\delta^2 C}{\delta p^i(x)\,\delta p^j(y)}; \tag{54}$$

$$C = \int d^4x \Lambda[\mathbf{p}; \mathbf{p}_\mu, \boldsymbol{\alpha}_\mu](x) \tag{55}$$

are decoupled. The requirement that $[\Delta_p^{-1}](k) = k^2 - m_\pi^2$ determines β:

$$\beta\frac{m_\varrho^2}{m_\varrho^2 + \beta\lambda^2 f_\pi^2} = 1$$

i.e.,

$$\beta = \frac{m_\varrho^2}{m_\varrho^2 - \lambda^2 f_\pi^2}. \tag{56}$$

We may therefore write

$$\boldsymbol{\alpha}'_\mu = \boldsymbol{\alpha}_\mu + \frac{\lambda f_\pi}{m_\varrho^2} D_\mu \mathbf{p}. \tag{57}$$

The axial meson propagator (53) is given by

$$[\Delta_{\alpha'}^{-1}]_{\mu\gamma}(k) = \alpha(k_\mu k_\nu - k^2 g_{\mu\nu}) + g_{\mu\nu}(m_\varrho^2 + \beta\lambda^2 f_\pi^2)^2.$$

The parameter λ is the value of the irreducible $\varrho\pi\pi$ vertex at all external momenta $= 0$:

$$\Lambda_{\varrho\pi\pi}(x) = \lambda[\boldsymbol{\varrho}_\mu \cdot \mathbf{p} \times \partial^\mu \mathbf{p}](x) + 0(\xi^3).$$

Estimates of this value indicate that

$$\lambda^2 \simeq \frac{1}{2}\left(\frac{m_\varrho}{f_\pi}\right)^2$$

so that

$$\beta \simeq 2.$$

If we assume that $\Delta_{\alpha'}$ is also dominated by a resonance, in addition to the previous assumption $\alpha = 1$, we obtain the mass of this resonance $m_{\alpha'}$:

$$(m_{\alpha'})^2 = m_\varrho^2 + \beta\lambda^2 f_\pi^2 \simeq 2m_\varrho^2.$$

The value $m_{\alpha'} \simeq \sqrt{2m_\varrho}$ is very close to the energy of the resonance Al ($J^P = 1^+$, $T^G = 1^-$, $\simeq 1070$ Mev).

The propagator for $\boldsymbol{\alpha}_\mu$, Δ_α, is given by

$$[\Delta_\alpha]_{\mu\nu}(k^2) = -\left(g_{\mu\nu} - \frac{1}{m_{\alpha'}^2} k_\mu k_\nu\right)\frac{1}{k^2 - m_{\alpha'}^2} - \left(\frac{\lambda f_\pi}{m_\varrho^2}\right)^2 k_\mu k_\nu \frac{1}{k^2 - m_\pi^2} + 0(\xi^4, \xi^2\varepsilon, \varepsilon^2). \tag{58}$$

The proof for this is left as an exercise.

Bibliography

The presentation here is essentially an elaboration of the work of B. Zumino,

1. B. Zumino, in *Proceedings of Trieste Conference on Renormalization* (1969, unpublished), *Lectures at Seminar on Electromagnetic Interactions and Vector Meson Dominance* (Dubna 1969, unpublished). *Lectures at Brandeis Summer Institute*, **1970** (MIT Press, 1970).

What is accomplished here is a justification at the phenomenological level of the results of the so-called field algebra, which is an attempt to unify the concept of vector meson dominance and current algebra in a field theoretical framework. For the development of field algebra, the reader is referred to

2. N. Kroll, T. D. Lee and B. Zumino, *Phys. Rev.*, **157**, 1376 (1967).
3. T. D. Lee, S. Weinberg and B. Zumino, *Phys. Rev. Letters*, **18**, 1029 (1967).
4. T. D. Lee and B. Zumino, *Phys. Rev.*, **163**, 1667 (1967).

We have not discussed all the ramifications of the phenomenological Lagrangian including vector and axial vector mesons such as nonminimal couplings, effects of vector and axial vector mesons on radiative corrections, etc. For these topics, see for example,

5. B. W. Lee and H. T. Nieh, *Phys. Rev.*, **166**, 1507 (1968).

For discussions on Schwinger terms, T*-products, and sea gull terms, see

6. D. Gross and R. Jackiw, *Nuclear Physics* **B14**, 269 (1969).
7. Wu Ki Tung, *Phys. Rev.*, **188**, 2404 (1969).

and references cited therein.

XII Chiral $SU(3) \times SU(3)$

12a Tensor analysis of $SU(3)$, $SU(3) \times SU(3)$

Before discussing the physics of the chiral $SU(3) \times SU(3)$, let us briefly review the tensor analyses of $SU(3)$ and $SU(3) \times SU(3)$.

The eight 3×3 traceless hermitian matrices of Gell-Mann, $\lambda_i = 1, 2, \ldots, 8$ normalized by

$$Tr\lambda_i\lambda_j = 2\delta_{ij} \tag{1}$$

are closed under commutation:

$$[\lambda_i, \lambda_j] = 2if_{ijk}\lambda_k, \tag{2}$$

where the structure constants f_{ijk} are completely antisymmetric in three indices and real. In Gell-Mann's basis $I_3 \sim \lambda_3/2$ and $Y \sim (1/\sqrt{3})\,\lambda_8$. The sets $(\lambda_1/2, \lambda_2/2, \lambda_3/2)$ and $[\lambda_6/2, \lambda_7/2, -(\lambda_3 - \sqrt{3}\lambda_8/2)]$ form the I spin and the U spin, respectively.

It is convenient to add

$$\lambda_0 = \sqrt{\frac{2}{3}}\,1 \tag{3}$$

to the set of eight traceless matrices λ_i. Equations (1) and (2) are true with λ_0 included, with $f_{0ij} = 0$. It is convenient to define the d-symbols by

$$\{\lambda_i, \lambda_j\} = 2d_{ijk}\lambda_k, \tag{4}$$

where d_{ijk} are completely symmetric in three indices and real. In particular

$$d_{0ij} = \sqrt{\frac{2}{3}}\,\delta_{ij}. \tag{5}$$

The generators of $SU(3) \times SU(3)$, Q_i, Q_i^5:

$$Q_i = \int d^3x V_i^0(x), \quad Q_i^5 = \int d^3x A_i^0(x), \tag{6}$$

where V_i^μ and A_i^μ are the octets of the vector and axial vector currents, satisfy the commutation relations:

$$\begin{aligned}
[Q_i, Q_j] &= if_{ijk}Q_k, \\
[Q_i, Q_j^5] &= if_{ijk}Q_k^5, \\
[Q_i^5, Q_j^5] &= if_{ijk}Q_k.
\end{aligned} \tag{7}$$

The chiral charges Q_i^+ and Q_i^-

$$Q_{\mathrm{I}}^{\pm} = \tfrac{1}{2}(Q_i \pm Q_i^5) \tag{8}$$

form disjoint $SU(3)$'s.

We shall write an arbitrary element of $SU(3) \times SU(3)$ as

$$\begin{aligned}
U(a,b) &= \exp\{-[ia\cdot Q + ib\cdot Q^5]\} \\
&= \exp\{-i[(a+b)\cdot Q^+ + (a-b)\cdot Q^-]\},
\end{aligned} \tag{9}$$

where a and b are eight-component vectors. We shall denote by χ_α the three-component spinor transforming under the $[SU(3)]_+$ generated by Q_i^+; by $\chi_{\dot\alpha}$ the spinor of the $[SU(3)]_-$ generated by Q_i^-. Under the infinitesimal action of (9), these transform as

$$\begin{aligned}
\chi_\alpha &\to \chi_\alpha + i[(a+b)\cdot\lambda/2]_\alpha^\beta\,\chi_\beta, \\
\chi_{\dot\alpha} &\to \chi_{\dot\alpha} + i[(a-b)\cdot\lambda/2]_{\dot\alpha}^{\dot\beta}\,\chi_{\dot\beta}.
\end{aligned} \tag{10}$$

In $SU(3)$, χ_α and its complex conjugate $\chi^\alpha = (\chi_\alpha)^*$ are not equivalent. Since $\lambda_i = \lambda_i^\dagger$, we have

$$\begin{aligned}
\chi^\alpha &\to \chi^\alpha + i(a+b)\cdot[-\lambda^T/2]_\beta^\alpha\,\chi^\beta \\
&= \chi^\alpha - i(a+b)\cdot\chi^\beta(\lambda/2)_\beta^\alpha
\end{aligned} \tag{11}$$

etc. All the irreducible representation of $SU(3)$ can be constructed by taking direct products of χ^{α}'s and χ^{β}'s;

$$\chi^{\lambda\mu\nu\cdots}_{\alpha\beta\gamma\cdots} \sim \chi^{\lambda} \otimes \chi^{\mu} \otimes \chi^{\nu} \cdots \otimes \chi_{\alpha} \otimes \chi_{\beta} \otimes \chi_{\gamma} \cdots$$

and symmetrizing with respect to upper indices, and with respect to lower indices, and removing the trace. Thus the irreducible tensor $\chi^{\lambda\mu\nu\cdots}_{\alpha\beta\gamma\cdots}$ satisfies

$$\varepsilon_{\sigma\lambda\mu}\chi^{\lambda\mu\nu\cdots}_{\alpha\beta\gamma\cdots} = \varepsilon^{\sigma\alpha\beta}\chi^{\lambda\mu\nu\cdots}_{\alpha\beta\gamma\cdots} = \chi^{\sigma\mu\nu\cdots}_{\sigma\alpha\beta\cdots} = 0.$$

Note that the symbols $\delta^{\alpha}_{\lambda}$, $\varepsilon_{\varrho\sigma\tau}$ are $SU(3)$ invariant and

$$\chi_{\varrho} \sim \varepsilon_{\varrho\sigma\tau}\chi^{\sigma\tau}.$$

The dimensionality of the irreducible tensor with n upper indices and m lower indices is

$$d = \left(1 + \frac{n + m}{2}\right)(1 + n)(1 + m).$$

The irreducible representation $[d, d']$ of $SU(3) \times SU(3)$ is the direct product of the irreducible representation d of $[SU(3)]_+$ and the irreducible representation d' of $[SU(3)]_-$.

Of particular interest in the ensuing discussions is the $[3, \bar{3}] \oplus [\bar{3}, 3]$ representation (18 dimensional):

$$M^{\dot{\beta}}_{a} \oplus M^{\beta}_{\dot{\alpha}},$$

which transforms under the infinitesimal transformations of Eq. (9) as

$$M^{\dot{\beta}}_{\alpha} \rightarrow M^{\dot{\beta}}_{\alpha} + i\frac{a}{2}\cdot[\lambda, M]^{\dot{\beta}}_{\alpha} + i\frac{b}{2}\cdot\{\lambda, M\}^{\dot{\beta}}_{\alpha}; \tag{12}$$

$$M^{\beta}_{\dot{\alpha}} \rightarrow M^{\beta}_{\dot{\alpha}} + i\frac{a}{2}\cdot[\lambda, M]^{\beta}_{\dot{\alpha}} - i\frac{b}{2}\cdot\{\lambda, M\}^{\beta}_{\dot{\alpha}}.$$

It is convenient to write

$$M^{\dot{\beta}}_{\alpha} \equiv (M)_{\alpha\beta} = \frac{1}{\sqrt{2}}[\lambda]_{\alpha\beta}\cdot(s + ip)$$

and

$$M^{\beta}_{\dot{\alpha}} \equiv (M^{\dagger})_{\alpha\beta} = \frac{1}{\sqrt{2}}[\lambda]_{\alpha\beta}\cdot(s - ip) \tag{13}$$

where s and p are real, nine-component vectors,

$$\lambda \cdot s = \sum_{i=0}^{8} \lambda_i s_i,$$

etc. and transform under infinitesimal chiral transformations as

$$\begin{aligned} s_i &\to s_i - f_{ijk} a_j s_k - d_{ijk} b_j p_k, \\ p_i &\to p_i - f_{ijk} a_j p_j + d_{ijk} b_j s_k. \end{aligned} \tag{14}$$

Note that

$$\frac{1}{\sqrt{2}} \lambda \cdot p$$

$$= \begin{pmatrix} \frac{\pi^0}{\sqrt{2}} + \frac{\eta_8}{\sqrt{6}} + \frac{\eta_0}{\sqrt{3}} & \pi^+ & K^+ \\ \pi^- & -\frac{\pi^0}{\sqrt{2}} + \frac{\eta_8}{\sqrt{6}} + \frac{\eta_0}{\sqrt{3}} & K^0 \\ K^- & \bar{K}^0 & -\sqrt{\frac{2}{3}}\eta_8 + \frac{\eta_0}{\sqrt{3}} \end{pmatrix}$$

in the conventional notation.

Under the operation of parity P, $Q_i^+ \leftrightarrow Q_i^-$, so that

$$P: \quad M(\mathbf{x}, t) \leftrightarrow \eta_p M^\dagger(-\mathbf{x}, t) \tag{15}$$

where η_p is the intrinsic parity of M. Choosing it to be $+1$, we have

$$P: \quad s(\mathbf{x}, t) \to s(-\mathbf{x}, t); \quad p(\mathbf{x}, t) \to -p(-\mathbf{x}, t). \tag{16}$$

In general, s and p have opposite parities.

The $SU(3) \times SU(3)$ invariants that can be constructed out of M and $M^\dagger$ are

1) $I_1 = TrMM^\dagger$,

2) $I_2 = TrMM^\dagger MM^\dagger$,

3) $I_0 = \det M$, $I_0^* = \det M^\dagger$. (17)

The other invariants such as $I_n = Tr(MM^\dagger)^n$, $n > 2$, are not independent of I_1, I_2, I_0 and I_0^*. The maximal groups that leave invariant each of these are 0(18) for I_1; $U(3) \times U(3)$ for I_2 and $SU(3) \times SU(3)$ for I_3.

The following simple facts about arbitrary 3×3 matrices will prove useful:

$$(1)\ \det M = \tfrac{1}{3} TrM^3 - \tfrac{1}{2}(TrM)(TrM^2) + \tfrac{1}{6}(TrM)^3$$
$$= M^3 - M^2 TrM + \tfrac{1}{2} M[(TrM)^2 - TrM^2]. \tag{18}$$

The last form is due to S. Coleman [*Hadrons and their Interactions*, edited by A. Zichichi, Academic Press (1968)].

(2) Any 3×3 matrix M may be written as

$$M = UXV^\dagger e^{i\phi}, \tag{19}$$

where $UU^\dagger = VV^\dagger = 1$, $\det U = \det V = 1$; $\phi = \frac{1}{3} \arg \det M$; X is diagonal with nonnegative real eigenvalues.

Proof Since $M^\dagger M$ is a positive semidefinite hermitian matrix, i.e.,

$$u^\dagger M^\dagger M u \geqq 0$$

for all 3 dimensional complex vectors u, we can write it as

$$M^\dagger M = VX^2V^\dagger,$$

where $VV^\dagger = 1$, $\det V = 1$ and X^2 is of the form

$$(X^2)_{\alpha\beta} = x_\alpha^2 \delta_{\alpha,\beta}$$

with $x_\alpha \geqq 0$. Define X to be

$$(X)_{\alpha\beta} = x_\alpha \delta_{\alpha,\beta}.$$

Now, let

$$M = WXV^\dagger.$$

Since

$$x_\alpha x_\beta (W^\dagger W)_{\alpha\beta} = x_\alpha x_\beta \delta_{\alpha\beta},$$

we may choose

$$W^\dagger W = 1$$

even if some of x_α's are zero. Now

$$\det W = (\det M)(x_1 x_2 x_3)^{-1}.$$

Therefore one may write $W = Ue^{i\phi}$; $UU^\dagger = 1$, $\det U = 1$, $3\phi = \arg\det M$.

(3) With $M = s + ip$, $s^\dagger = s$; $p^\dagger = p$, we have [Lévy, Nuovo Cimento **52***A*, 23 (1967)]

$$3(\det M + \det M^\dagger) = 2\{s^3\} - 3\{s\}\{s^2\} + \{s\}^3 - 6\{sp^2\} + 6\{sp\}\{p\} + 3\{s\}\{p^2\} - 3\{s\}\{p\}^2,$$

$$I_2 = Tr(MM^\dagger)^2 = \{p^4 + s^4 + 4p^2s^2 - 2(sp)^2\},$$

$$I_1 = Tr(MM^\dagger) = \{s^2 + p^2\}$$

where $\{\ \}$ denotes trace. Note that $Tr(MM^\dagger)^n$ can be expressible in terms of I_0, I_0^*, I_1 and I_2. For example

$$Tr(MM^\dagger)^3 = Tr(MM^\dagger)\,[\tfrac{3}{2}Tr(MM^\dagger)^2 - \tfrac{1}{2}[Tr(MM^\dagger)]^2] + 3\det MM^\dagger$$

$$= I_1[\tfrac{3}{2}I_2 - \tfrac{1}{2}(I_1)^2] + I_0I_0^*.$$

12b Broken chiral $SU(3) \times SU(3)$; Goldstone mode

The approximate invariance under $SU(3)$ of strong interactions and the approximate invariance under the chiral $SU(2) \times SU(2)$ of the non-strange world suggest that perhaps the chiral $SU(3) \times SU(3)$ is the global symmetry of the strongly interacting particles.

We have argued that the chiral $SU(2) \times SU(2)$ is weakly broken, and we have considered the ratio of the pion mass to a typical hadron mass (e.g., m_σ, m_ϱ or m_N) as an indication of the weakness of the symmetry breaking. Measured by the same yardstick, the breaking of the chiral $SU(3) \times SU(3)$ is by no means weak. We may conceive of the Goldstone mode of the Chiral $SU(3) \times SU(3)$ symmetry in which the kaons and the η-meson, as well as the pions are zero mass Goldstone bosons. In the real world, the kaon (or η) mass, however, is not small compared to the ϱ-mass, say. We have no *a priori* reason to believe that a first order perturbation theory around the Goldstone mode of the $SU(3) \times SU(3)$ will give a reliable approximation to the real world. However, we shall try this, if only to show that the scheme is not a tenable one.

How is the chiral $SU(3) \times SU(3)$ broken? A generalization of our ideas on the breaking of the chiral $SU(2) \times SU(2)$ suggests that the symmetry breaking Lagrangian is proportional to a scalar density which transforms like $[3, \bar{3}] \oplus [\bar{3}, 3]$:

$$\mathscr{L}' = \alpha_0 s_0 + \alpha_8 s_8 = Tr\gamma s, \tag{1}$$

where

$$s = \frac{1}{\sqrt{2}} \sum_{i=0}^{8} \lambda_i s_i$$

and

$$\gamma = \frac{1}{\sqrt{2}} (\lambda_0 \alpha_0 + \lambda_8 \alpha_8).$$

The symmetry breaking Lagrangian (1) will break the chiral $SU(3) \times SU(3)$ down to the isospin $SU(2) \otimes$ hypercharge $U(1)$, unless the parameters α_0 and α_8 are constrained in certain ways. For example if $\alpha_8 = 0$, then the $SU(3)$ is still an exact symmetry.

Under the axial transformations s transforms like

$$\delta s = -\{\tfrac{1}{2} b \cdot \lambda, p\} \tag{2}$$

so that the symmetry breaking term transforms like

$$-\frac{\delta \mathscr{L}'}{\delta b_i} = \frac{1}{2} Tr(\gamma \{\lambda_i, p\}). \tag{3}$$

We shall write γ as

$$\gamma = \gamma_1 + \gamma_2, \tag{4}$$

where

$$\gamma_1 = -\frac{\alpha_8}{\sqrt{2}} (\lambda_0 - \sqrt{2}\lambda_8) = -\sqrt{3}\alpha_8 \begin{pmatrix} 0 & & \\ & 0 & \\ & & 1 \end{pmatrix} \tag{5}$$

and

$$\gamma_2 = (\alpha_0 + \alpha_8/\sqrt{2}) \lambda_0 = \frac{\alpha_8 + \sqrt{2}\alpha_0}{\sqrt{3}} \begin{pmatrix} 1 & & \\ & 1 & \\ & & 1 \end{pmatrix}. \tag{6}$$

If $\gamma_2 = 0$, $\mathscr{L}'$ is invariant under the chiral $SU(2) \times SU(2)$, but $\mathscr{L}'$ breaks the $SU(3)$: If $\gamma_1 = 0$, then $\mathscr{L}'$ is invariant under the $SU(3)$, but $\mathscr{L}'$ breaks the chiral $SU(2) \times SU(2)$. This situation is depicted in

the diagram

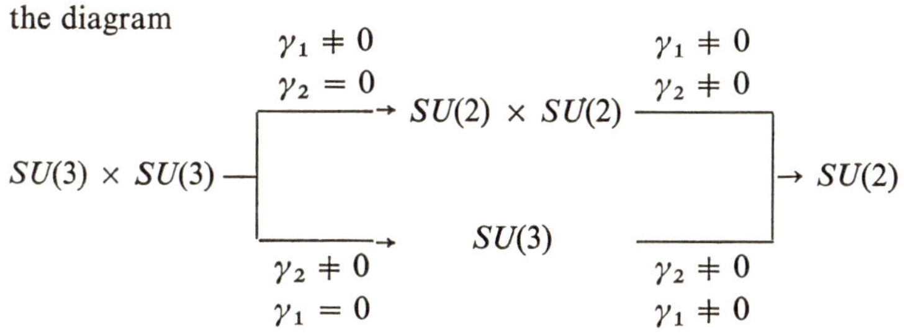

If $|\gamma_1| \gg |\gamma_2| \simeq 0$, the chiral $SU(2) \times SU(2)$ is weakly broken; in this case our previous considerations based on the weak breaking of the chiral $SU(2) \times SU(2)$ are justified: this scheme of the symmetry breaking was particularly emphasized by Gell-Mann, Oakes and Renner.

Let us consider some possible Goldstone modes of the $SU(3) \times SU(3)$ symmetry when $\gamma = 0$. Equation (9, 18) of the preceeding section:

$$[\Delta^{-1}(0)]_{ij}\,(T_a)_{jk}\,v_k = 0, \tag{7}$$

where $i, j, k = 1, \ldots, 18$ label the states of the $[3, \bar{3}] \oplus [\bar{3}, 3]$ representation and T_a, $a = 1, \ldots, 16$ are the representations of the generators, suggests the following cases of possible physical interest. The first is that the maximum invariance of the vacuum is the $SU(3)$, so that $v_0 = \langle s_0 \rangle_0 \neq 0$ while other v's vanish. In this case the octet of p.s. mesons are zero mass Goldstone bosons. The second possibility of physical interest is that the maximal invariance of the physical vacuum is the hypercharge-isospin $U(2)$. In this case $v_0 = \langle s_0 \rangle_0$ and $v_8 = \langle s_8 \rangle_0$ are nonvanishing; the zero mass Goldstone bosons are the octet of p.s. mesons and the quartet of scalar mesons (the hypothetical $\varkappa$ of $S = \pm 1$ and $T = \frac{1}{2}$). The lack of hard evidence for the existence of the $\varkappa$-mesons despite a long search makes me believe that this possibility is unlikely.

If the first possibility is indeed realized in nature, there is a question as to whether an additional discrete symmetry which is a subgroup of $SU(3) \times SU(3)$ is also present. Consider the $[3, \bar{3}] \oplus [\bar{3}, 3]$ representations of the $SU(3) \times SU(3)$ generators:

$$Q_i \sim \lambda_i/2; \qquad Q_i^5 \sim \gamma^5 \lambda_i/2.$$

It is easy to see that

$$Z \equiv \exp i \frac{4\pi}{\sqrt{3}} Q_8^5 \equiv \exp i\, 2\pi Y^5 \sim \exp i \frac{2\pi}{\sqrt{3}} \gamma_5 \lambda_8$$

commutes with all λ_i. In general

$$[Z, Q_i] = 0.$$

The elements 1, Z, Z^2, the parity $P(\sim\gamma_0)$, ZP and Z^2P all commute with Q_i and form S_3, the symmetric group on three objects. The question is whether the vacuum is invariant under S_3. For the things we shall discuss, it doesn't matter. If S_3 is an invariance, certain baryonic states appear in parity doublets. We shall not discuss this matter any further but refer the interested to Dashen's original discussion [*Phys. Rev.*, **183**, 1245 (1969)].

12c Enthalpy functional for chiral $SU(3) \times SU(3)$

The enthalpy functional of the 18*c*-fields, *s* and *p*, to order ξ^2, takes the form

$$A[s, p] = A^{(0)}[s, p] + A^{(2)}[s, p] \tag{1}$$

where $A^{(0)}[s, p]$ is a local functional of the invariants:

$$A^{(0)}[s, p] = \int d^4x \Lambda^{(0)}(x),$$
$$\Lambda^{(0)}(x) = \Lambda^{(0)}[I_0, I_0^*, I_1, I_2](x) \tag{2}$$

and $A^{(2)}[s, p]$ may be written as

$$A^{(2)}[s, p] = \int d^4x \Lambda^{(2)}(x),$$
$$\Lambda^{(2)}(x) = \{Tr(\partial_\mu M\, \partial^\mu M^\dagger)\, F_1[I_0, I_0^*, I_1, I_2]$$
$$+ Tr(\partial_\mu M\, \partial^\mu M^\dagger M M^\dagger)\, F_2[I_0, I_0^*, I_1, I_2]$$
$$+ \cdots\}(x). \tag{3}$$

As we have discussed in Sec. VII, the vacuum expectation values v are determined from the condition (7.36), or equivalently from

$$\left.\frac{\partial L(M, M^\dagger)}{\partial M_{\beta\alpha}}\right|_{M=v,\ M^\dagger=v^\dagger} = 0, \tag{4a}$$

$$\left.\frac{\partial L(M, M^\dagger)}{\partial (M^\dagger)_{\beta\alpha}}\right|_{M=v,\ M^\dagger=v^\dagger} = 0 \tag{4b}$$

where

$$L(M, M^\dagger) = \Lambda^0(M, M^\dagger) + \tfrac{1}{2}Tr\gamma(M + M^\dagger). \tag{5}$$

(In Eq. (5), $L(M, M^\dagger)$ and $\Lambda^0(M, M^\dagger)$ are to be considered as functions of numerical matrices M, $M^\dagger$) subject to the requirement that the masses associated with physical excitations must be non-negative:

$$\begin{aligned} -[\Delta_S(0)]_{ij}^{-1} &\equiv -\{\partial L(s,p)/\partial s_i\, \partial s_j\}\big|_{M=v, M^\dagger=v^\dagger} \geqq 0, \\ -[\Delta_p(0)]_{ij}^{-1} &\equiv -\{\partial L(s,p)/\partial p_i\, \partial p_j\}\big|_{M=v, M^\dagger=v^\dagger} \geqq 0, \\ & -\{\partial L(s,p)/\partial p_i\, \partial s_j\}\big|_{M=v,\ M^\dagger=v^\dagger} \geqq 0. \end{aligned} \tag{6}$$

The last of Eq. (6) is satisfied by the equality if the parity is conserved. Equations (4a), (4b), and (6) imply that the vacuum expectation values v, and $v^\dagger$ are given by the values of M and $M^\dagger$ which minimize the function $L(M, M^\dagger)$. We have not shown that v and $v^\dagger$ must be the *absolute* minimum of L. Nevertheless, let us assume that this is the case, so that v and $v^\dagger$ are determined uniquely in terms of γ from Eqs. (4a), (4b), and (6).

It is not guaranteed, *a priori*, that the minimum of L occurs at a real matrix $v = v^\dagger$. However, if $v \neq v^\dagger$ (while $\gamma = \gamma^\dagger$, by assumption), then the vacuum is not invariant under parity. We shall assume below that the dynamics is such that the minimum of L occurs at a real matrix v. Under this assumption, Eq. (4) may be expressed as

$$[\gamma]_{\alpha\beta} = Av_{\alpha\beta} + B[v^3]_{\alpha\beta} + C[v^{-1}]_{\alpha\beta} \det v$$

where

$$\begin{aligned} A &= -[\partial\Lambda^0/\partial I_1]_{M=M^\dagger=v}, \\ B &= -\tfrac{1}{2}[\partial\Lambda^0/\partial I_2]_{M=M^\dagger=v}, \\ C &= -[\partial\Lambda^0/\partial I_0]_{M=M^\dagger=v}. \end{aligned} \tag{7}$$

The parameters A, B, and C are constants depending on the eigenvalues of v. Since v is hermitian, it can be diagonalized in Eq. (7). Then Eq. (7) tells us that v and γ may be diagonalized simultaneously.

The solution v to order ε^0 (i.e., γ^0) is a solution of the matrix equation

$$0 = Av + Bv^3 + Cv^{-1} \det v. \tag{8}$$

Since we want a solution which preserves the hypercharge isospin $[U(2)]_I$, let us assume the solution of Eq. (8) which satisfies the stability conditions,

$$\Delta_s^{-1}(0) \leqq 0, \quad \Delta_p^{-1}(0) \leqq 0 \tag{9}$$

is unique, and is of the form of

$$v[\gamma = 0] = f\begin{pmatrix} 1 & & \\ & 1 & \\ & & \omega \end{pmatrix}. \tag{10}$$

We shall consider several ramifications:

(1) $w = 1$: $v[0] = \sqrt{\tfrac{3}{2}}f\lambda_0$, so that $v_0 = \sqrt{\tfrac{3}{2}}f$ and all other v's $= 0$. The equations (see Eqs. (12b, 7))

$$[\Delta_s^{-1}(0)]_{ij}\,(Q_a)_{jk}v_k = 0,$$
$$[\Delta_p^{-1}(0)]_{ij}\,(Q_a^5)_{jk}v_k = 0, \tag{11}$$

where $i, j, k = 0, 1, 2, \ldots, 8$ and $a = 1, 2, \ldots, 8$ tell us in what states Goldstone bosons appear: The first of Eq. (11) contains no useful physical information. The second states that the octet of p.s. mesons are Goldstone bosons. The symmetry of the vacuum is $SU(3)$.

(2) $\omega \neq 1, 0$, or -1: in this case v is a linear combination of λ_0 and λ_8, and

$$v_8 \neq 0, \quad \sqrt{2}v_0, \quad \text{or} \quad 2\sqrt{2}v_0.$$

The symmetry of the vacuum is $[U(2)]_I$. The first of Eq. (11) says that the hypothetical $\varkappa$-mesons have zero mass; the second says the octet of p.s. mesons is massless.

The other two cases, $\omega = 0$ and -1, have also been investigated by Bardeen and Lee. The case $\omega = 0$ does not seem to be physically relevant. The case $\omega = -1$ is equivalent to $\omega = +1$ after an $SU(3) \times SU(3)$ transformation, and need not be discussed.

We shall concentrate on case (1). We shall assume that, in the limit $\gamma \to 0$, the symmetry of $SU(3) \times SU(3)$ is manifest in the Goldstone mode in which the octet of p.s. mesons becomes massless. We will now construct the generating functional of the irreducible vertices of the p.s. meson octet, to order ξ^2 and γ. We have expressed our reservation about the perturbative treatment in powers of γ. Nevertheless, we shall do this, to show that the irreducible vertices of the meson octet are unique to this order.

The constraint equations, from which s_i, $i = 0, 1, \ldots, 8$ and p_0 must be determined in terms of p_i, $i = 1, 2, \ldots, 8$, are, to order ξ^0 and γ^0,

$$\frac{\delta_A^{(0)}[s, p]}{\delta s_i(x)} = 0 \quad \text{for} \quad i = 0, 1, \ldots, 8$$

and

$$\frac{\delta_A^{(0)}[s, p]}{\delta p_0(x)} = 0. \tag{12}$$

The two equations in (12) are solved if we demand

$$A'M + B'MM^\dagger M + C'^* M^{\dagger -1} \det M^\dagger = 0 \tag{13}$$

and its hermitian conjugate, where

$$\begin{aligned} A' &= -\partial \Lambda^{(0)}/\partial I_1, \\ B' &= -\tfrac{1}{2}\partial \Lambda^{(0)}/\partial I_2, \\ C' &= -\partial \Lambda^{(0)}/\partial I_0. \end{aligned} \tag{14}$$

A', B', and C' are functions of the invariants I_0, I_0^*, I_1, and I_2 and

$$[A', B', C']_{M=M^\dagger=v} = [A, B, C]. \tag{15}$$

Now, let us write M in the way suggested by Eq. (12a, 19):

$$M = UXV^\dagger e^{i\phi}, \tag{16}$$

where U and V are unitary unimodular and X is diagonal and positive semidefinite. From Eqs. (5) and (16), we infer that Eq. (13) and its hermitian conjugate can be solved if we constrain M such that

$$\begin{aligned} I_0 &= \det M = \det v = f^3, \\ I_1 &= TrMM^\dagger = Trv^2 = 3f^2, \\ I_2 &= Tr(MM^\dagger)^2 = Trv^4 = 3f^4. \end{aligned} \tag{17}$$

and

The first of Eq. (17) gives

$$\phi = 0, \qquad \det X = \det v$$

and the last two imply

$$TrX^2 = Trv^2,$$

$$TrX^4 = Trv^4.$$

The desired solution is

$$X = v = f\,\mathbf{1}. \tag{18}$$

(This solution for X ensures that $M \to v$ as $p \to 0$.)

We can therefore write M as

$$M = f e^{i\pi/f} = s + ip \tag{19}$$

where UV^+ is written as $\exp i\pi f^{-1}$,

$$\pi = \sum_{i=1}^{8} \lambda_i \pi_i, \qquad Tr\pi = 0.$$

Since the nonlinear transformation $p \to \pi$ is canonical,

$$p_i = \frac{f}{2} Tr\lambda_i \sin(\pi f^{-1})$$

$$= \pi_i + 0(\pi^3)$$

we may invoke the invariance of the on-shell T matrix under the nonlinear canonical transformations of c-fields and treat the octet of π's as the variables in constructing the nonlinear phenomenological Lagrangian density $\Lambda[\pi](x)$. It is unique to order ξ^2 and γ, and takes the form

$$\Lambda[\pi](x) = Tr(\partial_\mu M\, \partial^\mu M^\dagger) + \tfrac{1}{2} Tr\gamma(M + M^\dagger) \tag{20}$$

where M is given by Eq. (19). We note further that

$$p_0 = \frac{f}{2} Tr\lambda_0 \sin(\pi f^{-1}),$$

$$= 0(\pi^3)$$

and

$$s_i = \frac{f}{2} Tr\lambda_i \cos(\pi f^{-1}),$$

$$= 0(\pi^2).$$

The nonlinear phenomenological Lagrangian can be constructed by the same method for case (2), this time in terms of c-fields corresponding to the octet of p.s. mesons and the quartet of the $\varkappa$-mesons. The matrix M can be written in this instance as

$$M = e^{i\pi} e^{i\varkappa} v\, e^{-i\varkappa} e^{i\pi}$$

where v is as given in Eq. (10) with $\omega \neq 1, 0$ or -1, and $\varkappa$ is the matrix

$$\varkappa = \begin{pmatrix} 0 & 0 & \varkappa^+ \\ 0 & 0 & \varkappa^0 \\ \bar{\varkappa}^- & \bar{\varkappa}^0 & 0 \end{pmatrix}$$

[The c-fields π_i and $\varkappa$ need be further renormalized in order that the inverse propagators for these particles come out in the desired form.] We shall leave the details to the reader as an exercise; discussions on this and related topics may be found in Bardeen and Lee, and Nieh and Tsao [*Phys. Rev.*, to be published].

12d Electromagnetic mass difference—Dashen's theorem

The electromagnetic mass difference within an isospin multiplet is a long standing problem which has attracted a considerable amount of attention since the early days of particle physics. Usually, the electromagnetic mass difference is attributed to the second order electromagnetic interaction

$$H_{em} = \frac{e^2}{2} \int d^4x \int d^4y D^{\mu\nu}(y)\, T^* \left\{ j_\mu\left(x + \frac{y}{2}\right) j_\nu\left(x - \frac{y}{2}\right)\right\}, \tag{1}$$

where j_μ is the electromagnetic current

$$j^\mu(x) = V_3^\mu(x) + \frac{1}{\sqrt{3}} V_8^\mu(x) \tag{2}$$

and $D_{\mu\nu}$ is the photon propagator. The symbol T^* means that quantities that follow be time ordered, and if necessary certain contact terms be added in order to make the quantity inside the bracket Lorentz covariant and divergenceless [see Sec. 11].

In the usual treatment, a complete set of states is inserted between the two currents and the contributions from low mass states, together with the contributions from associated "seagull" graphs (i.e., the contact terms necessitated by the T^* prescription) are evaluated. The electromagnetic Hamiltonian of Eq. (1) transforms like a mixture of $T = 0$, 1 and 2 objects [and like a mixture of 1, 8 and 27 representations of $SU(3)$]. The electromagnetic mass differences computed in this manner are found to be in good agreement with reality for the $\Delta T = 2$ mass differences (such as $\pi^+ - \pi^0$), but in poor agreement for the $\Delta T = 1$ cases (such as $p - n$, $K^+ - K^0$). One could think of many possible causes for this. For example, one may argue that the saturation of intermediate state contributions by a few low lying states is a good approximation for the $\Delta T = 2$ cases, but not for the $\Delta T = 1$ cases, as indeed some arguments based on the Regge asymptotic behavior of relevant amplitudes suggest [Harari, *Phys. Rev. Letters*, **17**, 1303 (1967)]. Coleman and Glashow suggested that the interaction of Eq. (1) induces a vacuum expectation value of a neutral scalar field of the $T = 1$ neutral member of an octet, and the interaction of hadrons with this field enhances selectively the $\Delta T = 1$ electromagnetic mass differences over the $\Delta T = 2$ ones.

The electromagnetic current j^μ is invariant under the $[U(2) \times U(2)]_U$ generated by

$$\left[Q_6, Q_7, \frac{1}{2}\left(\sqrt{3}\,Q_8 - Q_3\right)\right], \qquad Q_3 + \frac{1}{\sqrt{3}} Q_8 \tag{3}$$

and their chiral partners (i.e., $Q_i \to Q_i^5$). The invariance of H_{em} under the U-spin transformations yield a number of sum rules, such as

$$m^2(\pi^+) - m^2(\pi^-) = m^2(K^+) - m^2(K^0) + \frac{\sqrt{3}}{2} m^2_{\eta\pi}, \tag{4}$$

$$m(\Xi^-) - m(\Xi^0) = m(\Sigma^-) - m(\Sigma^+) + m(p) - m(n), \tag{5}$$

where $m^2_{\eta\pi}$ is the off-diagonal element of the mass matrix $[-\Delta_p^{-1}]$. Equation (5) is in good agreement with experiment.

Let us now discuss the $\pi^+ - \pi^0$ mass difference. The quantity:

$$\langle T^*(\pi^i(x)\, \pi^j(y)\, j_\mu(\xi)\, j_\nu(\eta))\rangle_0 \tag{6}$$

may be evaluated by the method of Sec. II correctly to order ξ^2 and ε. From this, the mass-shift of the pion may be evaluated by joining the currents j_μ and j_ν by the photon propagator and performing the requisite integrations. It is found that, in a version of calculation which constructs the above matrix element in terms of irreducible vertices of π, ϱ, and A 1, the mass shift of the π^0 is zero

$$\Delta m^2(\pi^0) = 0 \tag{7}$$

and the mass shift of the π^+ is, in the Goldstone limit,

$$\Delta m^2(\pi^+) = \frac{3\alpha}{4\pi} m_\varrho^2 (2 \ln 2), \qquad m_\pi^2 \to 0. \tag{8}$$

For finite pion mass, Eq. (7) is still true, but the π^+ mass shift is logarithmically divergent. The appearance of a divergence is not surprising, since our construction of Eq. (6) is meant to be a good approximation for small ξ^2, and does not reflect the correct high energy behavior of the amplitude. In order to compensate for this, and to obtain a finite result, we may regularize the photon propagator as

$$\frac{1}{k^2} \to \frac{1}{k^2} \frac{-M^2}{k^2 - M^2}.$$

The result of a long calculation gives

$$\delta m_{\pi^+}^2 = A + B,$$

$$A = \frac{3\alpha}{4\pi} m_\varrho^2 \int_0^\infty \frac{dx}{(1+x)^2} \frac{1}{(x+v)^2} \times \left\{ x^2 + \frac{1}{3}\left(x^2 + \frac{x}{2} + \frac{1}{16}\right)\left[\frac{2(1+4rx)}{1+\sqrt{1+4rx}} - 1\right]\right\},$$

$$B = \frac{3\alpha}{4\pi} m_\varrho^2 \int_0^\infty \frac{dx}{(1+x)^2}\left(\frac{1}{1+vx}\right)^2 f\left\{\frac{x}{2f} - \frac{x}{g}\left(\frac{x}{f} - \frac{x}{f+g}\right) - \frac{r}{6}\left[\left(\frac{x}{4r} + 2\left(\frac{r}{f}\right)\right)\left(\frac{f}{g}\frac{f}{f+g} - \frac{1}{2}\right) + 4r\left(\frac{x}{f}\right)^2\left(\frac{x}{g}\right)\frac{1}{f+g}\right]\right\}, \tag{9}$$

where

$$r = (m_\pi^2/m_\varrho^2),$$

$$v = (m_\varrho/M)^2,$$

$$f = x + 2 - r; \quad g = \sqrt{f^2 + 4rx}. \tag{10}$$

[Equation (9) is written in a form suitable for computer input: For derivations see, for example, Lee and Nieh, *Phys. Rev.*, **166**, 1507 (1968): we have assumed $m(A\ 1) = \sqrt{2}m_\varrho$.] For a reasonable value of $v = 0.04$ which corresponds to the cut off mass $M = 5m_\varrho$, we obtain

$$\delta m(\pi^+) = 4.63 \text{ Mev}$$

compared to the experimental value $m(\pi^+) - m(\pi^-) = 4.604 \pm 0.004$ Mev.

A similar calculation can be carried out for K^+ and K^0. The role $A\ 1$ had for the pion is now taken over by KA. We find

$$\delta m^2(K^0) = 0 \tag{11}$$

and with the approximations $m_\varrho = m_\omega = m_\phi, m(KA) = m(A\ 1) = \sqrt{2}m_\varrho$, $\delta m^2(K^+)$ is given by the same expression as $\delta m^2(\pi^+)$ with the change $r = (m_K/m_\varrho)^2$. For the same cutoff mass $M = 5m_\varrho$, we obtain

$$\delta m(K^+) = 2.11 \text{ Mev}$$

to be compared with the experimental value $m(K^+) - m(K^0) = -3.94 \pm 0.13$ Mev. Even the sign does not come out right for this case.

One might hope that that "tadpole" mechanism of Coleman and Glashow would cure this malaise, but the situation is not so simple due to a quite general arguement advanced by Dashen. We shall state it as

Dashen's theorem In the Goldstone mode of the chiral $SU(3) \times SU(3)$ symmetry in which the octet of p.s. mesons are massless Goldstone bosons, the electromagnetic mass shifts of the neutral members K^0 and π^0, and the off-diagonal mass $m_{\eta\pi}$ all vanish to order α provided that

$$[A_0^i(x), H_{em}(t)] = [Q_5^i(t), \mathscr{H}_{em}(x)], \tag{12}$$

where

$$H_{em}(t) = \int d^3x \mathscr{H}_{em}(x). \tag{13}$$

We shall prove the theorem in a moment, but let us first note that the theorem implies

$$m(K^+) - m(K^0) = m(\pi^+) - m(\pi^0) \tag{14}$$

as follows from Eq. (4), to order ε^0, so no electromagnetic mechanism would get us out of this embarrassment. We might argue that the $SU(3) \times SU(3)$ breaking would account for the departure of the observed values from Eq. (14). No doubt. However, this casts, as we shall see, a grave doubt on the premise (which we have been reluctant to accept) that the $SU(3) \times SU(3)$ symmetry breaking is small and may be treated as a small perturbation on the Goldstone limit.

We will now describe the original proof. Let A^i_μ be the axial vector current in the presence of electromagnetism and $(A^i_\mu)^0$ the same quantity in its absence. Then

$$\partial^\mu (A^i_\mu)^0 = 0 \tag{15}$$

and

$$\begin{aligned} \partial^\mu A^i_\mu(x) &= -i[Q^i_5(t), \mathscr{H}_{em}(x)], \\ &= -i[A^i_0(x), \mathscr{H}_{em}(t)] \end{aligned} \tag{16}$$

by assumption. Consider now the identity, as $k \to 0$

$$\begin{aligned} &k^\mu k^\nu \int d^4x\, e^{ik\cdot x} \langle T(A^i_\mu(x)\, A^j_\nu(0))\rangle_0 \\ &= \int d^4x\, d^{ik\cdot x} \langle T(\partial^\mu A^i_\mu(x)\, \partial^\nu A^j_\nu(0))\rangle_0 - i\langle [Q^j_5, [Q^i_5, \mathscr{H}_{em}]]\rangle_0 . \end{aligned} \tag{17}$$

As $k \to 0$, the left hand side vanishes. The last term on the right is of order α, so we need compute the first term to this order. The first term on the right is in general of order α^2 except the pole term which is of order α, since there is a denominator of order α in this case. Now consider $\langle 0 | \partial^\mu A^i_\mu | pj \rangle$, where (pj) is the j-th member of the p.s. meson octet carrying momentum p. We need compute this only to order α:

$$\langle 0 | \partial^\mu A^i_\mu(x) | pj \rangle \cong \frac{f}{2} (m^2)_{ij}\, e^{-ip\cdot x} .$$

Therefore, Eq. (17) is equivalent to, in the limit $k \to 0$,

$$\left(\frac{f}{2}\right)^2 \sum_{k,l} (m^2)_{ik} \left[\frac{-i}{m^2}\right]_{kl} (m^2)_{lj} = + i \langle [Q^j_5, [Q^i_5, \mathscr{H}_{em}]]\rangle_0$$

or

$$(m^2)_{ij} = \left(\frac{2}{f}\right)^2 \langle [Q_5^j, [Q_5^i, \mathscr{H}_{em}]] \rangle_0 . \tag{18}$$

For $i, j = 3, 6, 7$ and 8, the double commutator on the right vanishes and the theorem follows.

The proof just given implies that there cannot be a tadpole mechanism as envisaged by Coleman and Glashow in the Goldstone mode of the Chiral $SU(3) \times SU(3)$ limit, or that, if such a mechanism does exist, its contribution must be cancelled exactly by nontadpole contributions. We shall see that the latter is the case from the following proof of Dashen's theorem:

Let $U^{em}[s, p]$ be the generating functional of the s, p-irreducible vertices due to the second order (i.e., $e^2 \sim \alpha$) electromagnetic interaction. Since the e.m. interaction of Eq. (1) is invariant under the $[U(2) \times U(2)]_U$, it follows that

$$\frac{\delta U^{em}[s, p]}{\delta \beta_a} = 0 = \int d^4x \sum_{i=0}^{8} \left\{ \frac{\delta U^{em}}{\delta s_i(x)} \frac{\delta s_i(x)}{\delta \beta_a} + \frac{\delta U^{em}}{\delta p_i(x)} \frac{\delta p_i(x)}{\delta \beta_a} \right\}, \tag{19}$$

where β_a is the generating parameter of the axial U spin transformations, i.e., $a = 3, 6, 7$ and 8. Furthermore, from Eq. (12a, 14) we have

$$\frac{\delta s_i(x)}{\delta \beta_a} = -d_{aij} p_j, \qquad \frac{\delta p_i}{\delta \beta_a} = +d_{aij} s_j .$$

Now differentiating Eq. (19) with respect to p_j and setting

$$s_j = v_j = \sqrt{\frac{3}{2}}\, f d_{j0} \qquad p_j = 0$$

i.e., taking the Goldstone mode of the chiral $SU(3) \times SU(3)$ limit, we find

$$-\int d^4x \left. \frac{\delta U^{em}}{\delta p_a(x)\, \delta p_i(0)} \right|_{\substack{s=v \\ p=0}} = \frac{1}{f} d_{aij} \eta_j^{em} \tag{20}$$

or

$$-\{[\Delta_p^{em}(0)]^{-1}\}_{ai} = \frac{1}{f} d_{aij} \eta_j^{em}, \tag{21}$$

where $-[\Delta_p^{em}(0)]^{-1}$ is the nontadpole contribution to the inverse p.s. meson propagator due to H_{em}, so that

$$[\delta m_p^2(\text{nontadpole})]_{ij} = -\{[\Delta_p^{em}(0)]^{-1}\}_{ij} \tag{22}$$

and η_j^{em} is the effective source for the electromagnetic tadpole

$$-\eta_j^{em} = \left.\frac{\delta U^{em}}{\delta s_j}\right|_{\substack{s=v\\p=0}}. \tag{23}$$

The tadpole contribution to the electromagnetic mass is

$$-[\delta m_p^2(\text{tadpole})]_{ij} = \Gamma_{k;ij}(0:00)\,[i\Delta_s(0)]_{kl}\,[i\eta_l^{em}], \tag{24}$$

where $\Gamma_{k;ij}(p;q,-p-q)$ is the irreducible sp^2-vertex:

$$\frac{\delta^3 U}{\delta s_k(x)\,\delta p_i(y)\,\delta p_j(z)} \equiv \tilde{\Gamma}_{k;ij}(x;yz),$$

$$\int d^4x\,d^4y\,d^4z\,e^{ip\cdot x+iq\cdot y+iq'\cdot z}\,\tilde{\Gamma}_{k;ij}(x;yz)$$

$$= (2\pi)^4\,\delta^4(p+q+q')\,\Gamma_{k;ij}(p;qq'). \tag{25}$$

Now, from the invariance of $U[s,p]$ under axial transformations one establishes that

$$-\Gamma_{l;ij}(0;00)\,[\Delta_s(0)]_{lk} = \frac{1}{f}\,d_{kij}; \quad \begin{array}{l} i,j = 1,2,\ldots,8,\\ k = 0,1,\ldots,8. \end{array} \tag{26}$$

[The derivation of Eq. (24) is very similar to that of Eq. (20): it follows from

$$\left.\frac{\delta^3 U}{\delta s_i(x)\,\delta p_j(y)\,\delta\beta_k}\right|_{\substack{s=v\\p=0}} = 0, \qquad \begin{array}{l} k = 1,2,\ldots,8,\\ i,j, = 0,1,2,\ldots,8. \end{array}$$

The details are left as an exercise.] Combining Eqs. (24) and (26) we obtain

$$-[\delta m_p^2(\text{tadpole})]_{ij} = \frac{1}{f}\,d_{ijk}\eta_k^{em}. \tag{27}$$

Now, taking $i = a = 3, 6, 7$ or 8 in Eq. (27) and combining with Eqs. (21) and (22), we obtain

$$[\delta m_p^2]_{aj} = [\delta m_p^2(\text{tadpole})]_{aj} + [\delta m_p^2(\text{nontadpole})]_{aj} = 0$$

which proves Dashen's theorem.

Dashen's theorem has a far reaching consequence in its implication on the $SU(3) \times SU(3)$ symmetry breaking. One might argue that the departure of the physical mass splittings from the Dashen theorem is due to the $SU(3) \times SU(3)$ breaking. However, if indeed the observed mass differences are to find a quantitative explanation in the interplay between the electromagnetic interaction and the $SU(3) \times SU(3)$ breaking, then it implies that the terms of order $\varepsilon\alpha$ is as large as terms of order α. In that case, we have no reason to believe that terms of order $\varepsilon^2\alpha$, $\varepsilon^3\alpha$, ... may be neglected. We must conclude, therefore, that the strength of the $SU(3) \times SU(3)$ breaking is not small, and the perturbative treatment of this effect is, to say the least, unreliable.

Then how are the Gell-Mann Okubo formula and the like justified? This is an old question, raised first by Yang and Oakes in a different context. I do not have a pat answer, but the following possibilities suggest themselves.

1) In the symmetry breaking Lagrangian $\alpha_0 s_0 + \alpha_8 s_8$, α_8 is small compared to α_0. It is legitimate to treat $\alpha_8 s_8$ as s small perturbation, but not $\alpha_0 s_0$. This would explain the Gell-Mann Okubo formula, but would be in contradiction with our views on the chiral $SU(2) \times SU(2)$ symmetry and its breaking.

2) The parameter which characterizes the $SU(3)$ breaking α_8 is much larger than that for the chiral $SU(2) \times SU(2)$ breaking $\alpha_8 + \sqrt{2}\alpha_0$. Higher order terms in α_8 are important, but somehow, only the octet parts of these terms are enhanced relative to other parts. This is the old idea of the octet enhancement.

3) There is nothing wrong with the perturbative approach to the chiral $SU(3) \times SU(3)$ breaking. Rather, the form of the symmetry breaking we accepted (the so-called Gell-Mann, Oakes, Renner model) is incorrect. In particular there is a nonelectromagnetic, isospin violating interaction. This would explain the mass differences within isomultiplets and the puzzle in the η decay (see bibliography). This possibility has been discussed vigorously in connection with the divergence problem in weak interactions (Gatto and collaborators; Cabibbo and his school), or in conjunction with the explanation for the Cabibbo angle (Cabibbo and Maiani; Oakes) or the *CP* violation (Pati). This is something of a radical departure from the conventional thinking, but it is a worthy idea that should not be taken lightly.

12e Speculation on broken chiral $SU(3) \times SU(3)$

The above discussion is based on the assumption that in the exact $SU(3) \times SU(3)$ limit, the symmetry manifests itself *in the Goldstone mode*. It is possible that this assumption is false.

We can imagine that the exact $SU(3) \times SU(3)$ limit of the world is a Wigner mode in which the octets of scalar and pseudoscalar mesons are completely degenerate and have finite masses; when the symmetry of the Lagrangian is reduced from the $SU(3) \times SU(3)$ to the $SU(2) \times SU(2)$, the pions and only the pions become the massless Goldstone bosons:

$SU(3) \times SU(3)$ ⟶	$SU(2) \times SU(2)$
Wigner mode (octets of scalar and p.s. mesons degenerate)	Goldstone mode (pions are massless Goldstone bosons).

In a renormalized $SU(3) \times SU(3)$ σ-model, such a situation can easily be realized. Consider for example the $SU(3) \times SU(3)$ symmetric Lagrangian

$$\mathscr{L}_s = \frac{1}{2} Tr\, \partial_\mu M\, \partial^\mu M^\dagger - \frac{\mu^2}{2} TrMM^\dagger$$

$$- \lambda Tr(MM^\dagger)^2 - \varrho(TrMM^\dagger)^2 - 3\nu(\det M + \det M^\dagger).$$

It can be shown easily that the positivity of the Hamiltonian requires

$$\lambda + \varrho \geqq 0, \qquad \varrho \geqq 0$$

and that if the condition

$$\mu^2 \geqq \frac{9}{4} \frac{\nu^2}{\lambda + 3\varrho}$$

is satisfied, the symmetry of the vacuum is that of $SU(3) \times SU(3)$, and the octet of scalar and pseudoscalar mesons will have a finite and degenerate mass.

Now consider the symmetry breaking term of the form

$$\mathscr{L}_b = Tr\gamma_1 s, \qquad \gamma_1 = \gamma \begin{pmatrix} 0 & & \\ & 0 & \\ & & 1 \end{pmatrix}.$$

The symmetry of the total Lagrangian $\mathscr{L}_s + \mathscr{L}_b$ is $SU(2) \times SU(2)$. As γ_1 increases from zero the degenerate 18 mesons split up into 5 groups corresponding to the $SU(2) \times SU(2)$ contents of the $[3, \bar{3}] + [\bar{3}, 3]$ representation of $SU(3) \times SU(3)$: $[\frac{1}{2}, \frac{1}{2}]$ consisting of π and σ, $[\frac{1}{2}, 0] + [0, \frac{1}{2}]$ consisting of K and $\varkappa$, $[0, 0]$ of a scalar iso-singlet, $[0, 0]$ of a pseudoscalar iso-singlet, and another $[\frac{1}{2}, \frac{1}{2}]$ consisting of a scalar isotriplet and a pseudoscalar isosinglet. It is not difficult to verify that the group consisting of π and σ has the lowest mass among the five. For μ^2 in the range (putting $\lambda = 0$ for simplicity)

$$\frac{3}{4}\frac{\nu^2}{\varrho} \leqq \mu^2 \leqq \frac{9}{4}\frac{\nu^2}{\varrho},$$

there is a critical value of $\gamma = \gamma_c$ at which the masses of π and σ become zero. Beyond this value, the σ field develops a nonvanishing vacuum expectation value; the symmetry of the vacuum is reduced to the $[U(2)]_I$ and the pions become zero mass Goldstone particles. With the additional small breaking of $SU(2) \times SU(2)$, such a picture accounts for the exceptionally small pion mass. The various theorems that apply to a small departure from the Goldstone mode holds only for $SU(2) \times SU(2)$ and pions, but not for $SU(3) \times SU(3)$ and the octet of pseudoscalar mesons. (In the language of current algebra, we may say that the soft K and η meson limits do not exist.)

We have yet to work out all the consequences of this hypothesis, but the picture seems to me very appealing, since it explains why the $SU(2) \times SU(2)$ symmetry in the Goldstone mode is a good approximation to nature, but not the Goldstone mode of $SU(3) \times SU(3)$. Why the Gell-Mann Okubo formula should be good in such a scheme is a complete mystery to me at the moment, however.

Bibliography

The following sources were often consulted in formulating this chapter:

1. M. Lévy, *Nuovo Cimento*, **52**, 23 (1967).
2. W. A. Bardeen and B. W. Lee, *Phys. Rev.*, **177**, 2389 (1969).
3. R. Dashen, *Phys. Rev.*, **183**, 1245 (1969).
4. N. Cabibbo and L. Maiani, *Phys. Rev.*, **D1**, 707 (1970).
5. M. Gell-Mann, R. Oakes and B. Renner, *Phys. Rev.*, **175**, 2195 (1968).

The second proof of Dashen's theorem given in the text is a generalization of a remark of Cabibbo and Maiani (loc. cit.).

The conventional treatments of the $\eta \rightarrow 3\pi$ based on the chiral $SU(3) \times SU(3)$ *and* the weakly broken $SU(3)$ gives a rate for this process which is much too small compared to experiment. The situation is reviewed in

6. J. S. Bell and D. G. Sutherland, *Nuclear Physics*, **B4**, 315 (1968).

A cure for this, in terms of an isospin violating nonelectromagnetic interaction is discussed by a number of authors; for example, see

7. K. Wilson, *Phys. Rev.*, **179**, 1499 (1969),
8. R. Oakes, *Phys. Letters*, **30B**, 262 (1969).

The experimental situation is not yet completely clear, but the predictions of the chiral $SU(3) \times SU(3)$ [based on the Gell-Mann, Oakes, Renner philosophy] on the K_{l_3} form factors may be contradicted by experiment. See for the review:

9. M. K. Gaillard and L. M. Chounet, K_{l_3} *Form Factors* CERN 70-14, May 1970 (unpublished).

There is a recent preprint

10. R. Brandt and G. Preparata, *Ann. Phys.* (New York) to be published

which discusses the so-called weak *PCAC*. The weak *PCAC* attributes the success of *PCAC* to the pion pole dominance in on-shell amplitudes, and rejects the view expounded here. I do not agree with these authors, but it is nevertheless an interesting article and the workers in this field are well advised to familiarize themselves with it.

The implementation of the idea of the dynamical octet enhancement will continue to be an important problem in particle theory, especially in conjunction with the chiral $SU(3) \times SU(3)$ breaking. See in particular Parts IX and X of

11. M. Gell-Mann and Y. Ne'eman, *The Eightfold Way*, W. A. Benjamin, Inc., New York (1964).